Contents

Examining
GCSE Human Biology

Morton Jenkins
Head of Science, Howardian High School, Cardiff
Joint Chief Examiner in GCSE Biology, WJEC
Member of the working parties for National Criteria
and Grade Related Criteria for Biology
GCSE Subject Leader for Biology, WJEC

Hutchinson

London · Melbourne · Auckland · Johannesburg

Hutchinson Education

An imprint of Century Hutchinson Ltd
62–65 Chandos Place, London WC2N 4NW

Century Hutchinson Australia Pty Ltd
PO Box 496, 16–22 Church Street, Hawthorn,
Melbourne, Victoria 3122, Australia

Century Hutchinson New Zealand Ltd
PO Box 40-086, Glenfield, Auckland 10, New Zealand

Century Hutchinson South Africa (Pty) Ltd
PO Box 337, Bergvlei 2012, South Africa

First published 1987 by Century Hutchinson Ltd
Reprinted 1987

© Morton Jenkins 1987

Illustrations © Century Hutchinson Ltd 1987

Designed, illustrated and typeset in 11/12 Plantin by 𝐅\ Tek-Art Ltd, Croydon

Printed in Great Britain by Butler & Tanner Ltd, Frome

British Library Cataloguing in Publication Data

Jenkins, Morton
    Examining GCSE human biology,
    1. Human physiology
    I. Title
    612   QP36

ISBN 0-09-164731-2

Acknowledgements   I am grateful to Mr. Alan Jenkins for his ideas for cartoons and to Dr. J.A.F.
Napier of the Welsh Regional Blood Transfusion Centre for help with the section
on blood. I also express my thanks to Pat Rowlinson and Ruth Holmes for their
invaluable editorial help. Last, but not least, I am grateful to my wife Betty and
sons Alan and Andrew who tolerated the intrusion into family life admirably
while the book was being written.

Acknowledgements are due to the following for permission to reproduce
photographs:
Alan Beaumont p. 11; Bettman Archive/BBC Hulton Picture Library p. 152;
Biophoto Associates pp. 20, 184; Botanical Society of the British Isles p. 86;
Camera Press Ltd. pp. 10, 13, 60, 80, 84, 87, 111, 117, 121, 175, 197; J. Allan
Cash Ltd. pp. 9, 16, 27, 83, 85, 86, 101, 102, 105, 106, 107, 143, 179; FAO p.
103; Glaxochem p. 48; Sally and Richard Greenhill pp. 47, 72, 73, 119, 152, 197,
205, 229; Robert Harding Picture Library Ltd. p. 202; Health Education Council
p. 168; Mansell Collection pp. 46, 70, 83, 118, 137, 141; Mencap p. 211;
Mothercare pp. 91, 204; Picturepoint Ltd. p. 106; Queen Charlotte's Hospital
p. 10; Ann Ronan Picture Library p. 142; St. Bartholomew's Hospital pp. 54,
169; Science Photo Library p. 203; Alan Thomas pp. 8, 9, 10, 52, 55, 59, 62,
63, 64, 65, 67, 104, 105, 107, 108, 109, 112, 151, 176, 199; USDA pp. 102, 103;
C. James Webb pp. 18, 20, 21, 33, 34, 36, 38, 39, 43, 44, 45, 64, 91, 96, 110,
130, 132, 137, 163, 174, 177, 185, 211, 212, 220, 221, 222, 239, 240, 241;
Wellcome Foundation Ltd p. 49; WHO pp. 83, 84, 103, 115, 120, 141, 143;
World Bank Photo Library p. 121

The author and publishers are grateful to *The Guardian* and *New Scientist* for
permission to reproduce extracts in the book.

Preface

Before the development of GCSE syllabi it was possible to obtain an 'O' Level grade in twenty-five different ways for many subjects offered by examination boards. The users of examination certificates questioned the comparability of standards. For example, was a Grade C from Board X the same as a Grade C from Board Y? Could you be certain that someone with a Grade C knew and understood a particular body of facts? In order to answer these questions, the Secretary of State for Education invited GCE and CSE boards' Joint Council to submit recommendations for 16+ National Criteria.

It was recognized that there was a need to produce standardization for syllabi, with common aims, objectives, content and assessment patterns.

By September 1982, work had begun on writing the National Criteria. Early in 1983, a wide range of interested groups were invited to comment on the first drafts. They included schools, colleges, further education and higher education establishments, Schools' Council, HM Inspectorate, and certain professional associations. By March 1985, after certain modifications of the original drafts, the Secretary of State for Education and the Secretary of State for Wales approved the National Criteria. They believed that the goal of having guidelines for the GCSE, agreed nationally by all parties concerned, had been achieved.

In Biology (and Human Biology) the aims describe the educational purposes of following a course for the GCSE examination. They are:

1 To develop an interest in, and enjoyment of, the study of living organisms.
2 To encourage an attitude of curiosity and scientific enquiry.
3 To promote an appreciation of the importance of experimental and investigatory work in the study of Biology (and Human Biology).
4 To promote respect for all forms of life.
5 To develop knowledge and understanding of fundamental biological concepts and principles.
6 To develop an awareness of:
a relationships between living organisms;
b relationships between living organisms and their environment;
c the effect of human activities on these relationships.
7 To develop a range of manipulative and communicative skills appropriate to the subject.
8 To develop an ability to use these skills to identify and solve problems.
9 To promote an awareness and appreciation of the development and significance of biology in personal, social, economic and technological contexts.
10 To provide:
a a worthwhile educational experience for all, whether or not students are intending to study biology beyond GCSE level;
b a suitable preparation for careers which require a knowledge of biology;
c a suitable foundation for further studies in biology and related disciplines.

Objectives These reflect the aims that are measurable.
1 *Knowledge with understanding*
Candidates should be able to:
a demonstrate knowledge and understanding of biological facts and principles, practical techniques and safety precautions;

b demonstrate knowledge and understanding of the personal, social, economic and technological applications of biology in modern society;

c use appropriate terminology in demonstrating this knowledge.

2 *Skills and processes*

Candidates should be able to:

a make and record accurate observations;

b plan and conduct simple experiments to test given hypotheses;

c formulate hypotheses and design and conduct simple experiments to test them;

d make constructive criticisms of the design of experiments;

e analyse, interpret and draw inferences from a variety of forms of information including the results of experiments;

f apply biological knowledge and understanding to the solution of problems, including those of a personal, social, economic and technological nature;

g select and organize information relevant to particular ideas and communicate this information cogently in a variety of ways;

h present biological information coherently.

In assessing whether candidates have achieved these objectives, certain criteria will be used in assisting grading. These are called criteria related grades. Examination boards will have statements in their syllabi which clearly describe the performance expected for the award of certain grades. To gain a grade, candidates will have to demonstrate mastery of skills or competences described by the criteria related grades. The criteria are defined so that they:

a are expressed in positive terms and must reflect what the candidates know and can do (they are not expressed in terms of failure);

b refer to the positive attainments of candidates gaining A, C, and F grades in the GCSE examination which will grade candidates from A to G.

The least able candidates will be given credit for what they know, rather than be penalized for what they don't know. In an assessment, questions will be set by examination boards which will be able to be completed by the very good, the average, and below average. Motivation of pupils through positive achievement is one of the aims of the criteria related grades.

Why criteria related grades? For a number of years it has been recognized that the users of examination certificates wanted to know more than just a grade. This was particularly true of employers and admission officers for further education. In other words, a mere symbol, for example A, B or C, does not give enough information on the 'likely levels of competence' of a prospective employee or student. In 1983, educationalists began to look at the possibility of defining criteria which could be applied to the examination performance of candidates. The grade obtained should then reflect a degree of mastery of certain skills. For example, in all subjects which have 'Biology' in their title, the skills are under three 'Domain' headings:

1 Knowledge with understanding.

2 Handling information and solving problems.

3 Experimental skills and investigations.

Each candidate could have a certificate which could show not only an aggregate grade but also a grade in each of the domains. The user of the certificate could therefore have a more complete picture of the abilities of a candidate. But what do the 'Domains' mean?

Knowledge with understanding
Candidates should be able to demonstrate their knowledge and understanding of:
1.1 Biological terminology.
1.2 Biological facts, principles and concepts.
1.3 Practical techniques and safety precautions.
1.4 Everyday uses of biology.
1.5 Personal, social, economic, technological and environmental applications of biology.

Handling information and solving problems
Candidates should be able to demonstrate their ability, within a biological context to:

2.1 Locate, select, organize, translate and present information (diagrammatic graphical, numerical, written and oral).
2.2 Use information to draw inferences and report trends.
2.3 Use knowledge to present reasoned explanations for phenomena, patterns and relationships and to propose hypotheses.
2.4 Make predictions.
2.5 Solve problems including some of a quantitative nature.

Experimental skills and investigations
Candidates should be able to demonstrate their ability, within a biological context, to show:

3.1 Observational skills.
3.2 Procedural and manipulative skills.
3.3 Measurement skills.
    The expectations of a Grade F, C, and A candidate will be defined so that their performance in each domain can be measured.

For the student   This book attempts to treat Human Biology in a way which conforms to the National Criteria for Biology. The content of nationally agreed syllabi is divided into four themes:

Theme one: Diversity of organisms.
Theme two: Relationships between organisms and with the environment.
Theme three: Organization and maintenance of the individual.
Theme four: Development of organisms and the continuity of life.

    The text covers Human Biology by following these themes, bearing in mind the important consideration of the examination consortia to reduce descriptive content and to emphasize the skills and processes of understanding.
    At the end of each chapter, within the themes, are questions which assess the skills of knowledge with understanding, handling information and solving problems. The questions which do this are identified with the symbol I, for those covering knowledge with understanding, and II for those covering handling information and problem solving. Skills involving experimentation and investigation are encouraged to be developed through practical work where this is possible. Certain topics in Human Biology do not lend themselves to a practical treatment. In these cases, problem-solving exercises are set using second-hand data.

THEME 1 Diversity of organisms

1.1 Man the peculiar animal

What makes us different from non-living things?

Perhaps there is a single answer to this question. Man, like all living things, can take elements from outside his body and rearrange them to form parts of himself. For example, a girl and her pet kitten can eat the same cooked meat. In one case the meat is changed into human flesh and in the other it is turned into parts of a kitten. A plant can take in carbon dioxide from the air, water and minerals from the soil, and turn them into leaves, stems and roots. All materials taken in by organisms are made of atoms and these are the same as the atoms that make up our bodies. So what do the great variety of living things on our planet have in common?

All living things do the following:

1 *Feed*. One of the differences between plants and animals is their method of feeding (see Fig.1.1.1). Animals have to eat plants or they have to eat other animals which themselves have eaten plants. Plants can make their own food using carbon dioxide, water, minerals, energy from sunlight and their own special green chemical, chlorophyll. The process by which plants make their food is called **photosynthesis**.

Fig. 1.1.1 Feeding

2 *Release energy from glucose*. This is called **respiration** (see Fig.1.1.2). All the energy locked up in glucose is released if oxygen is used. Only some of the energy is released when oxygen is not used.

3 *Get rid of waste material*. When energy is released from glucose we are left with a waste material called carbon dioxide. When we take in more protein than we need, the extra cannot be stored and is changed into a waste material called urea. Both carbon dioxide and urea are removed from the body. This is called **excretion** (see Fig.1.1.3). Other waste materials such as minerals also have to be excreted.

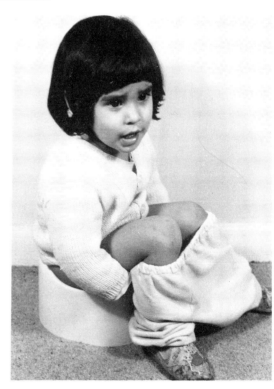

Fig. 1.1.2 Respiration

Fig. 1.1.3 Excretion

4   *Move.* Movement from one place to another takes place in all animals at some stage in their lives (see Fig.1.1.4). They have to do this to obtain food. Plants do not usually move from place to place but often move parts of themselves towards or away from light, water, and gravity.

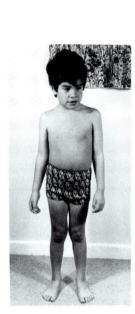

Fig. 1.1.5   Growth

Fig. 1.1.4   Movement

5   *Grow.* Growth involves an increase in size and only takes place when new materials are made in the body from materials that have been taken from the outside (see Fig.1.1.5).

6   *Increase in numbers.* This is called **reproduction** (see Fig.1.1.6). It makes sure that the various types of living things do not die out but continue to produce new generations of themselves.

7   *Sense changes in their surroundings.* This is called **sensitivity** (see Fig.1.1.7). Living organisms benefit by being able to react to changes in their surroundings, such as sound, light, chemicals and touch. Harmful changes can thus be avoided and helpful changes can be used to advantage.

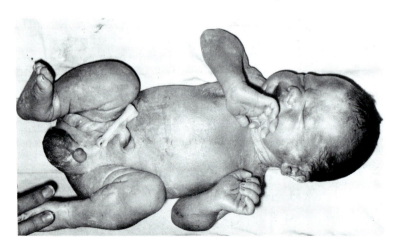

Fig. 1.1.6   Reproduction

Fig. 1.1.7   Sensitivity

What makes Man different from other animals?

Man is the only animal that can be ashamed of the fact that he is one. After all, it is not considered a compliment to call a person an animal! However, Man moves in search of his food and so, by definition, must be an animal. Most differences between Man and other animals are very obvious because he looks so different and behaves so differently from other animals. We can consider some of these differences in more detail.

Animals can be sorted into groups which have certain things in common. For example, there are the animals without backbones (**invertebrates**) as shown in Fig.1.1.8 and the animals with backbones (**vertebrates**) which include Man (see Fig.1.1.9).

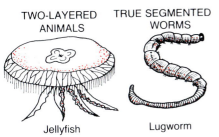

TWO-LAYERED ANIMALS — Jellyfish

TRUE SEGMENTED WORMS — Lugworm

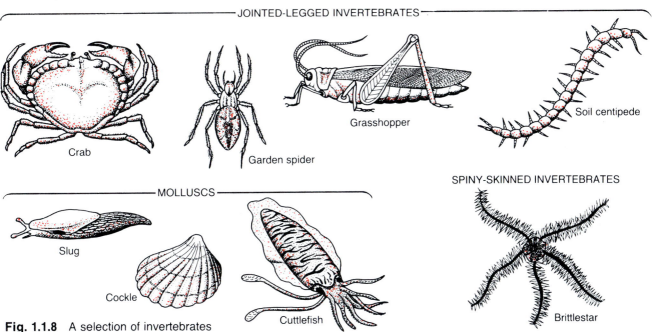

JOINTED-LEGGED INVERTEBRATES

Crab   Garden spider   Grasshopper   Soil centipede

MOLLUSCS

Slug   Cockle   Cuttlefish

SPINY-SKINNED INVERTEBRATES

Brittlestar

Fig. 1.1.8   A selection of invertebrates

Animals with backbones can be sorted into **classes**:

Class	Features
FISH	Body temperature is the same as the surroundings. Gills and paired fins are present.
AMPHIBIA	Body temperature is the same as the surroundings. They have slimy skins. They usually have limbs and lay eggs in water.
REPTILES	Body temperature is the same as the surroundings. They have scaly skins. They lay eggs on land.
BIRDS	Body temperature is constant. They have feathers.
MAMMALS	Body temperature is constant. They have hair and produce milk for their young.

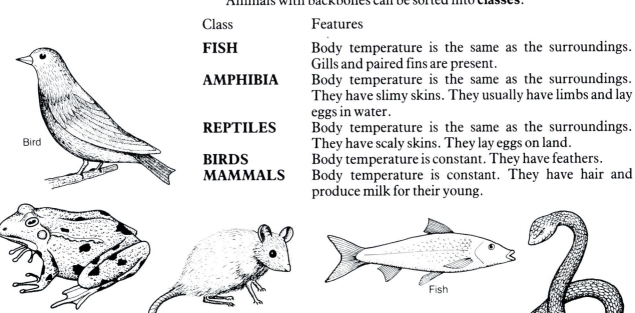

Bird   Amphibian   Mammal   Fish   Reptile

Fig. 1.1.9   The classes of vertebrates

What makes Man different from other mammals?

The class of vertebrates called mammals is further divided into **orders**, members of which have features in common (see Fig.1.1.10).

Main orders of mammals	Features
Monotremes	Mammals which lay eggs.
Marsupials	Typically have pouches in which the young develop.
Rodents	Chisel-shaped front teeth for continuous gnawing.
Bats	Front limbs modified as wings that can flap.
Carnivores	Flesh eating mammals with teeth modified for tearing.
Ungulates	Hoofed, grazing mammals.
Elephants	Elongated nose and upper lip forming a trunk.
Insectivores	Insect eaters with pointed surfaces to their teeth for crushing food.
Whales	Hairless, streamlined bodies, no hind-limbs, fore-limbs modified as flippers.
Primates	Highly developed brain, forward facing eyes. Hands modified for grasping. The young undergo long periods of growth, development and parental care.

MONOTREMES

Duck-billed platypus

MARSUPIALS

Kangaroo

RODENTS

Rat

BATS

Flying fox

CARNIVORES

Tiger

UNGULATES

Cow

ELEPHANTS

African elephant

INSECTIVORES

Mole

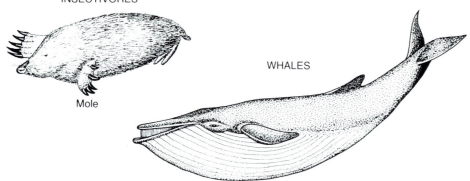

WHALES

Blue whale

PRIMATES

Chimpanzee

Fig. 1.1.10   The main orders of mammals

The evolution of Man

If our planet were to be visited by people from a different planet (see Fig.1.1.11), how would they describe Man as he is today? A computer analysis might read like this:

```
HE KILLS THE LARGEST LIVING LAND ANIMAL,
MAKES BALLS FROM ITS TEETH AND KNOCKS THEM
INTO HOLES IN A TABLE WITH STICKS. THEY
CALL IT SNOOKER (SEE FIG 1.1.12).
    HE TAKES THE GUT OF A SMALL ANIMAL,
STRETCHES IT ON A WOODEN FRAME AND KNOCKS
RUBBER BALLS BACKWARDS AND FORWARDS.
THEY CALL IT TENNIS (SEE FIG. 1.1.13).
    HE TAKES LEAVES FROM A PLANT, DRIES THEM,
ROLLS THEM INTO PAPER, PUTS THEM IN HIS
MOUTH AND SETS FIRE TO THEM. THEY CALL IT
SMOKING (SEE FIG 1.1.14).
    HIS TECHNOLOGY IS ADVANCED ENOUGH TO
CARRY HIM TO THE MOON IN SPACE SHIPS. THEY
CALL IT PROGRESS (SEE FIG. 1.1.15).
```

Fig. 1.1.11 Visitors from another planet

Fig. 1.1.13 'They call it tennis'

Fig. 1.1.12 'They call it snooker'

Fig. 1.1.14 'They call it smoking'

Fig. 1.1.15 'They call it progress'

Perhaps the first three of these examples may seem ridiculous to you when compared with the fourth but they serve to show the vast variety of Man's behaviour. The very complex brain of Man is capable of creating all of the above activities. This is what makes him different from all other life on Earth.

However, if people from a different planet had visited our planet several millions of years ago, they would have reported very different facts about Man. Perhaps their report would have been something like this:

```
HE IS COVERED WITH HAIR. HE KILLS ANIMALS
WITH STICKS AND EATS THEM (SEE FIG. 1.1.16).
COMPARED TO OTHER HAIRY ANIMALS, HIS TEETH
ARE NOT VERY SPECIAL. THEY ARE NOT VERY
POINTED FOR EATING MEAT OR VERY FLATTENED
FOR EATING PLANTS.
   HE SOMETIMES WALKS ON TWO LEGS, BUT
USUALLY CRAWLS ON ALL FOURS, USING HIS HANDS.
```

Fig. 1.1.16 'He kills animals with sticks and eats them'

Was this animal the same as Man as we know him today? It took fifteen million years for this ancestor of Man to develop or change into present-day Man. This process of very gradual change is called **evolution**. It has taken place throughout the history of Earth. Plants and animals have changed in order to survive changes in their surroundings. Those that did not change or adapt died out and thus became extinct.

An example of change is our climate which has altered during the millions of years since the Earth was first formed.

Some 350 million years ago, much of Britain was a tropical swamp and so was much of Antarctica. However, 500 000 years ago Britain was covered in ice. Today, these areas have a different climate and only those animals and plants that have changed to suit new conditions have survived. How do we know? Mainly from the evidence of fossils. These are the remains of plants and animals which have become preserved in rocks over millions of years. When animals and plants die their skeletons often remain and become covered by mud. During the course of time more and more mud settles on these remains. The vast pressure of all this mud helps to squeeze out water. The drying action of the climate and the chemical reactions between the mud and the skeletons cause minerals to form in the skeletons. This process eventually results in fossils being formed. Fossils of tropical plants and animals are to be found in Britain and the Antarctic.

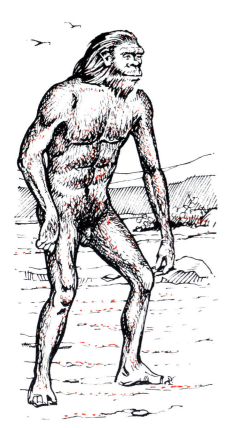

Of course, not all organisms that die change into fossils. Most are eaten by bacteria. In fact, the chances of organisms being fossilized are small. It depends on where they died. Those that died in swamps or in estuaries will have the best chance because they will be covered in mud quite quickly. Those animals that live in trees have the lowest chance of becoming fossils. They would usually fall to the ground and be eaten. The chances of finding fossils are even smaller but even so, the fossil record of evolution is very impressive. There is even a fossil record of Man's evolution together with collections of human bones which have not had time to change into fossils.

In 1891 a Dutch scientist, Dr Dubois, found some bones and a skull in Java. He saw that they were similar to a gorilla's bones but they were not exactly the same. Dubois realized that the bones were from a different animal which no one had known about before. He called it 'upright ape man' (*Pithecanthropus erectus*). Java Man was probably the earliest of our human ancestors, and would have been about 1.5m tall when fully grown (see Fig.1.1.17).

In a cave near Peking, in 1921, the bones of about forty humans were discovered. These Peking men had primitive tools with them and could have been cannibals. They lived about 500 000 years ago (see Fig.1.1.18).

Some 50 000 years ago lived a more human-like animal. He was Neanderthal Man (see Fig.1.1.19). Found with his bones were stone tools and weapons. Stone-Age Man had evolved.

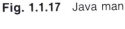

Fig. 1.1.17 Java man

Fig. 1.1.18 Peking man

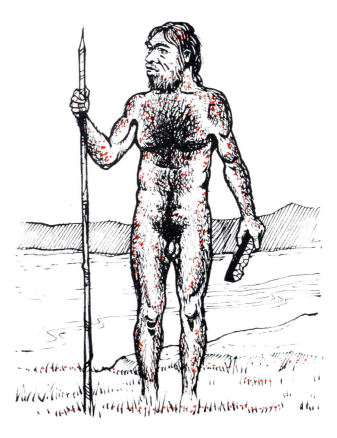

Fig. 1.1.19 Neanderthal man

Many discoveries have been made since Dubois' major find. These discoveries are like parts of a jig-saw puzzle which will eventually make up a complete picture of Man's evolution. He has so many features in common with the gorilla that it now seems certain that one of Man's close relatives is the gorilla (see Fig.1.1.20). However, Man is unique in having such a well-developed brain and the ability to think (see Fig.1.1.21). This has resulted in (1) his ability to use complex language; (2) the tendency for cooperation in groups towards a common aim and (3) the development of technology to control the environment, e.g. agriculture and machines.

Fig. 1.1.20   A close relative of man – the gorilla

Fig. 1.1.21   Rodin's 'The Thinker'

Questions: Domains I and II

1 The diagrams, **A**, **B**, **C** and **D** show the hands of some primates.

a Explain how the hands of **A**, **B**, and **C** are specialized for a particular function. (I)
b What is the importance of the opposability of the thumb shown in **D**? (I)

c The diagrams **E**, **F**, **G**, and **H** show the feet of the same types of primates. State three differences between the foot of H and those of the other primates. (II)
d How are these differences related to man's ability to walk erect? (I)

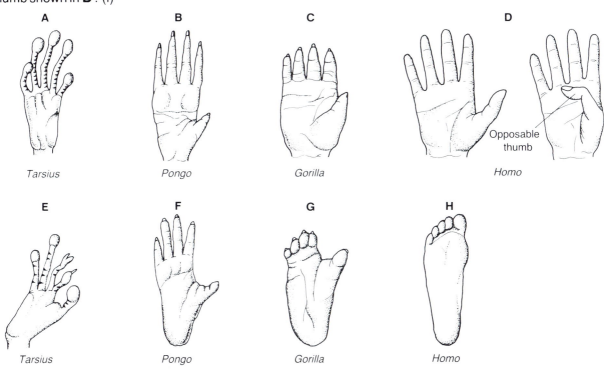

A — *Tarsius* B — *Pongo* C — *Gorilla* D — *Homo* — Opposable thumb

E — *Tarsius* F — *Pongo* G — *Gorilla* H — *Homo*

2 The skulls are of Man (A) and a gorilla (B).

a State five ways as seen in the illustrations in which the human skull differs in structure or form from that of the gorilla. (II)
b Comment on the significance of three of these differences. (I)

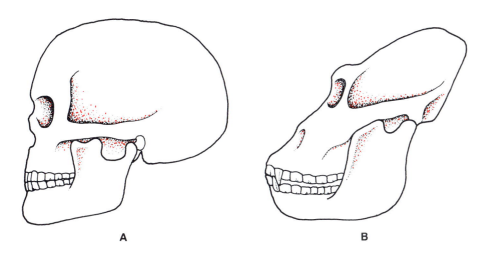

A B

1.2 What are we made of?

Cells

The building blocks of all living things are called **cells**. They are the basic units of life. Most living things are made of vast numbers of cells but there are some that are made of single cells. Cells are usually so small that school microscopes are not powerful enough to show their detailed structure. A photomicrograph (a photograph taken through a school microscope) of human cells which have been stained, looks like the one in Fig.1.2.1. Only three structures can be clearly seen:

1 The **nucleus** which controls all the activities of the cell by means of chemical instructions.

2 The **cytoplasm**. This is the substance of all living things. It is a jelly-like material consisting of water and dissolved substances made from carbon, hydrogen, oxygen, nitrogen, phosphorus and sulphur.

3 The **cell membrane**. This acts as a barrier to keep valuable materials inside the cell and to keep unwanted materials out. It also allows essential materials like water and food to pass in and waste products to pass out.

If more details of a cell are required, then it must be viewed at a greater magnification. This can be achieved with an electron microscope (see Fig.1.2.2). With this instrument scientists can see objects magnified hundreds of thousands of times. Fig.1.2.3 shows a diagram of a thin slice through a cell as revealed by the electron microscope. Note the presence of structures (**organelles**) which were not visible with a school microscope.

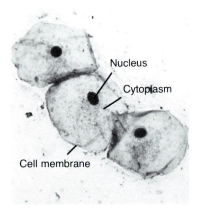

Fig. 1.2.1 Photomicrograph of three cheek cells

Fig. 1.2.2 An electron microscope in use

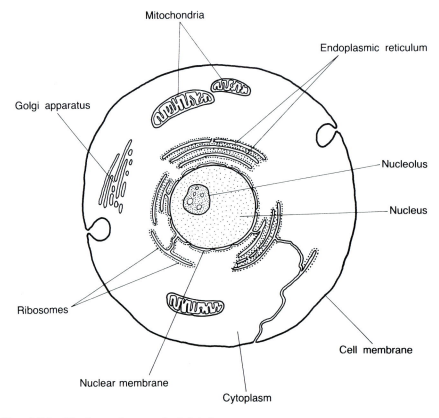

Fig. 1.2.3 Electron microscopical detail of an animal cell

1 **Mitochondria**. These contain chemicals which help the process of energy release from glucose.
2 **Endoplasmic reticulum**. This is a series of membranes. It allows materials to pass in and out of the nucleus and provides a large surface for the attachment of ribosomes.
3 **Ribosomes**. These are the sites of manufacture of protein.
4 **Golgi apparatus**. This is the packaging department of the cell. It forms envelopes made of membranes for transport of materials out of the cell.
5 **Nucleolus**. There may be more than one of these. They contain chemicals which are used in the manufacture of protein.
6 **Nuclear membrane**. This forms a boundary between the nucleus and the cytoplasm, but it allows a two-way movement between these parts due to the presence of pores.

All the cell's organelles act together in many complex ways and a vast number of chemical reactions take place inside them. The chemical reactions have at least one thing in common, that is that they all need special proteins called enzymes. An **enzyme** is a chemical produced by living things that speeds up the rate of chemical reactions. They:

a   only work within a narrow range of temperature;
b   are destroyed (denatured) when boiled;
c   only work within a narrow range of acidity and alkalinity;
d   only speed up specific chemical reactions: i.e. one particular enzyme will only be helpful to one particular reaction.

Tissues and organs

The basic structure of Man, in terms of materials, is similar to that of other animals. He has many specialized systems (see Fig.1.2.4). These are essential for carrying out all the characteristics of living organisms:

System		Function
Skeletal system	—	Support and movement.
Nervous system	—	Coordination and sensitivity.
Digestive system	—	Digestion and absorption.
Respiratory system	—	Exchange of gases.
Blood system	—	Transport and defence.
Excretory system	—	Elimination of waste.
Hormonal system	—	Chemical coordination.
Reproductive system	—	Production of reproductive cells and for the development of the embryo.

The various systems are made of organs, for example the heart, lungs, stomach and womb. Under Theme 3 (see p. 90) these will be considered in more detail. Here it is important to realize that organs are made of tissues. A tissue is a group of similar cells which work together to perform a particular function. Organs may be made of more than one type of tissue.

If we are to study the vast number of types of tissues of the body we can start by dividing them into:

1 **connective tissues** (tissues which support the body or connect parts together);

2 **epithelia** (singular **epithelium**) (tissues which cover the outside or inside of organs).

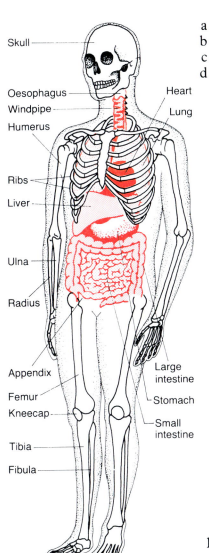

Skull
Oesophagus
Windpipe
Humerus
Ribs
Liver
Ulna
Radius
Appendix
Femur
Kneecap
Tibia
Fibula
Heart
Lung
Large intestine
Stomach
Small intestine

Fig. 1.2.4 Some of the main body systems

Connective tissues These are divided into the following groups:

a **Packing tissue** occurring between organs and under the skin. Its main function is to hold organs and organ systems in place.

b **Muscle** for movement of the body and for movements of organs. There are three main types:

i **Skeletal** (joined to skeleton), see Fig. 1.2.5.

ii **Smooth** (in many hollow organs), see Fig. 1.2.6.

iii **Cardiac** (in the heart), see Fig. 1.2.7.

All muscle cells can contract and relax when supplied with oxygen and a source of energy (glucose).

Fig. 1.2.5 Photomicrograph of skeletal muscle

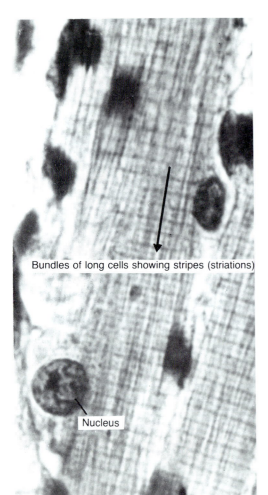

Fig. 1.2.6 Photomicrograph of smooth muscle

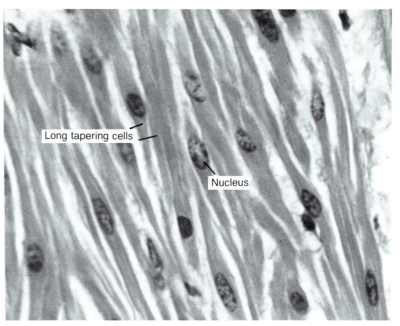

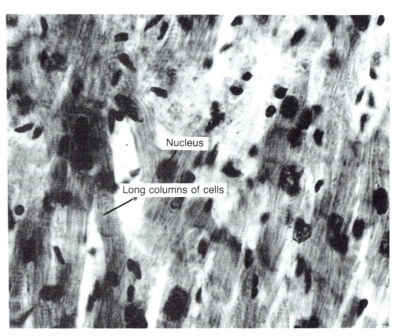

Fig. 1.2.7 Photomicrograph of cardiac muscle

c **Bone** and **cartilage**.
These are used for supporting the body or parts of it (see Figs.1.2.8 and 1.2.9).

d **Blood** (see p. 137).
This is unusual because it is a 'liquid' tissue, made of cells in a liquid matrix called **plasma** (see Fig.1.2.10).

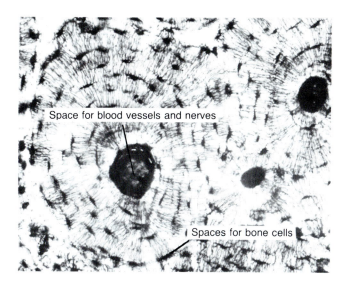

Fig. 1.2.8 Photomicrograph of a transverse section of bone (showing Harversian systems)

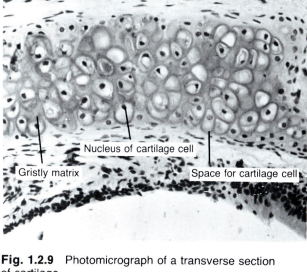

Fig. 1.2.9 Photomicrograph of a transverse section of cartilage

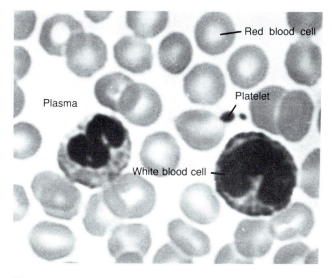

Fig. 1.2.10 Photomicrograph of a blood smear

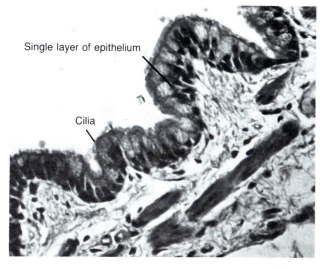

Fig. 1.2.11 Photomicrograph of ciliated epithelium

Epithelia These usually occur in sheets of one or more layers of cells. The cells can be flattened, cubical, or column shaped. Some of the cells can make and secrete (pour out) certain chemicals, e.g. digestive enzymes (see p. 129). Others may have small hair-like structures called **cilia** (singular cilium) which beat backwards and forwards (see Fig.1.2.11). The cilia often line organ systems which need to be kept clean or need to pass currents of material along. The respiratory system is built in this way. Dust particles are trapped in the cilia and are carried to the throat. In this way they are removed and prevented from interfering with exchanging gases. Parts of the reproductive system are lined with cilia which help to carry reproductive cells along (see p. 218).

Questions: Domains I and II

1 The table shows the rate of activity of an enzyme at different temperatures.

Temp °C	0	5	10	15	20	25	30	35	40	45	50	55	60
Rate (mg of product per minute)	1.8	2.4	3.7	4.9	7.4	9.3	13.4	17.2	19.0	19.0	8.1	1.7	0

a Draw a graph of these results. (II)
b State the optimum temperature for this enzyme. (II)
c Explain the rate of activity at 17°C. (I)
d Explain the results at temperatures above 45°C. (I)
e Name two factors other than temperature which would affect the rate of enzyme activity. (I)

Experimental skills and investigations: Domain III

1 The investigation of the action of the enzyme, catalase

Catalase is an enzyme found in all cells. It breaks down the chemical, hydrogen peroxide which is formed in all cells as a waste product. The formula for hydrogen peroxide is H_2O_2.

Materials
Fresh liver from the butcher, hydrogen peroxide, three test tubes, a splint for testing gases, Bunsen burner, tongs, labels.

Procedure
1 Take two pieces of liver, each about 1 cm^3. Place one in a test tube half full of water and boil it.
2 Carefully pour hydrogen peroxide into each of two test tubes until they are about half full. Label them A and B.
3 To A add a piece of unboiled liver. Test any gas given off with a glowing splint.
4 Repeat stage 3 with test tube B and the boiled liver.
5 Describe your observations.
6 Explain your observations.
7 What was the purpose of using boiled liver in tube B?
8 Suggest why it is necessary to break down hydrogen peroxide in living cells.

2 The investigation of a property of the cell membrane

An artificial cell membrane can be made from a substance called Visking tubing. It only allows molecules to pass through if they are small enough. The formula for a molecule of sugar is $C_{12}H_{22}O_{11}$. The formula for water is H_2O.

Materials
Visking tubing, capillary tubing, beaker, retort stand, 10% sugar solution, mm rule.

Procedure
1 Take a 10cm length of Visking tubing. Wet it under a tap to make it easier to use.
2 Tie a knot in the bottom of the Visking tube.
3 Half fill the Visking tube with 10% sugar solution.
4 Now set up the apparatus as shown in the diagram.
5 Mark the level of the sugar solution in the capillary tube. Mark the level of the water in the beaker.
6 Measure the levels of the sugar solution and water at 10 minute intervals.
7 Plot a graph of height in capillary tube against time, over a period of one hour.
8 Explain your results.
9 What would you expect to happen to living human cells if they were placed in a strong sugar solution?

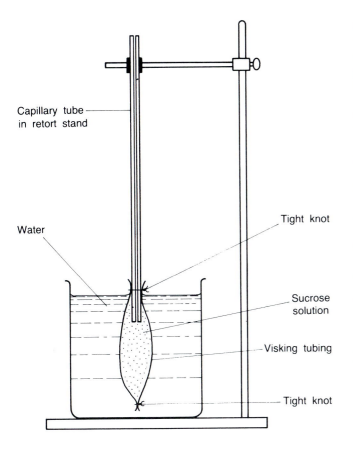

Capillary tube in retort stand

Water

Tight knot

Sucrose solution

Visking tubing

Tight knot

Materials
1% urea, 0.5% urease, two corked specimen tubes, test tube, Bunsen burner, Universal Indicator paper, paper clip, syringes for measuring volumes of liquids, Universal Indicator colour charts for pH.

Procedure
1 Set up two specimen tubes A and B like the one in the diagram.
2 Boil $2\,cm^3$ of 0.5% urease for two minutes in a test tube.
3 Add $2\,cm^3$ unboiled 0.5% urease and $2\,cm^3$ 1% urea to tube A.
4 Add $2\,cm^3$ boiled 0.5% urease and $2\,cm^3$ 1% urea to tube B.
5 Replace the corks so that the specimen tubes are as shown in the diagram.
6 Record the colour of the indicator paper in both tubes at 3-minute intervals over a period of 15 minutes.
7 Use the colour chart to find the pH value of (a) 1% urea, (b) water, (c) the contents of each tube (i) at the beginning and (ii) after 15 minutes.
8 Explain your observations.
9 What evidence is there to suggest that urease is an enzyme?

3 An investigation into the action of urease on urea

Urea is a substance produced by living organisms. It can be changed into ammonium carbonate to prevent it harming living cells.

$$UREA + WATER \xrightarrow{\text{urease}} AMMONIUM\ CARBONATE$$

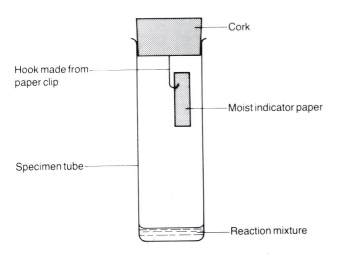

Cork

Hook made from paper clip

Moist indicator paper

Specimen tube

Reaction mixture

THEME 2 Relationships between organisms and with the environment: human interaction with the environment

2.1 View from a pyramid

Food chains

Man is dependent on green plants. He eats them directly or he eats animals which have eaten plants. It is possible to trace the diet of Man back to green plants by means of a **food chain** (see Fig. 2.1.1).

GRASS → COW → MAN

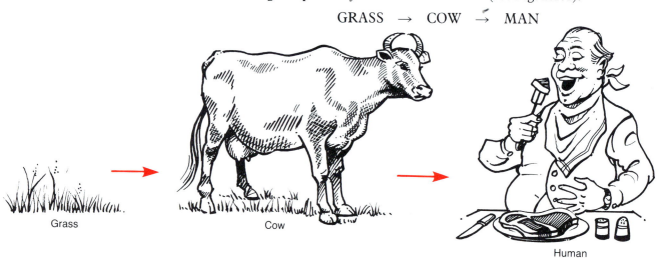

Grass   Cow   Human

Fig. 2.1.1 A simple food chain

The arrows not only show that the cow eats grass and that Man eats the cow. They also show the direction of the flow of energy. Energy from sunlight is changed to chemical energy in grass by photosynthesis. When the cow eats the grass this chemical energy is then changed into a form which can be used in the flesh of the cow. When Man eats the cow he changes this energy into chemical energy which he can use in his own tissues.

Green plants produce food by photosynthesis and so are called **PRODUCERS**. These are consumed by **first CONSUMERS** – herbivores. Herbivores are then consumed by **second consumers** – carnivores. (See Fig. 2.1.2.)

Food chains may have four or more links in them e.g.

GRASS → APHID → LADYBIRD → FROG → TROUT → MAN

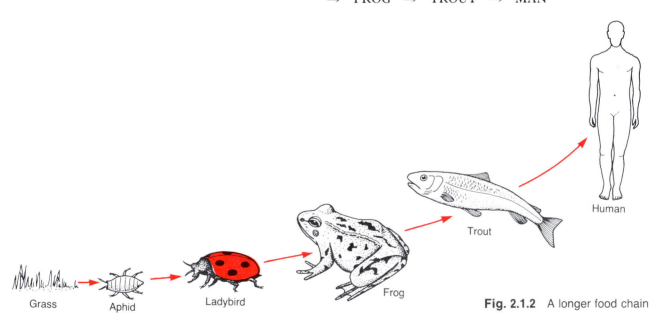

Fig. 2.1.2 A longer food chain

Food webs

Of course, the feeding habits of Man are too complex to be represented by a single food chain. How many people, for instance, rely on a constant supply of trout for their food? A more realistic picture of Man's feeding habits can be built into a **food web**. This is a series of inter-connected food chains such as the one shown in Fig. 2.1.3.

Between each link of a food chain there is considerable loss of energy. This is due to: (1) heat lost during respiration; (2) chemical energy loss due to excretion; and (3) chemical energy loss due to death and decay. A very small fraction of the energy from sunlight finds its way to Man at the end of a food chain.

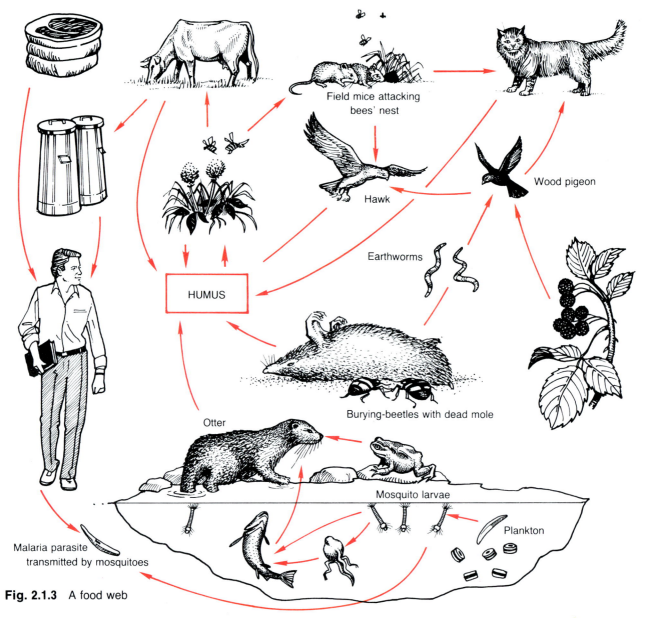

Field mice attacking
bees' nest

Hawk

Wood pigeon

Earthworms

HUMUS

Burying-beetles with dead mole

Otter

Mosquito larvae

Plankton

Malaria parasite
transmitted by mosquitoes

Fig. 2.1.3   A food web

Food pyramids   It follows then, that an enormous amount of producer is needed to provide enough energy for one person. If Man were to eat grass then there would not be so much energy loss. However, Man is not adapted to feed on a diet solely of grass. The mass of producer needed to support Man can be shown in the form of a **pyramid** (see Fig. 2.1.4). As energy passes from producer to carnivore, there is less of it available to each feeding level.

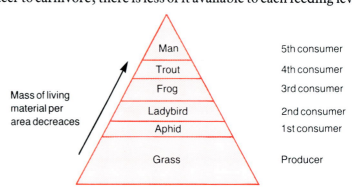

Mass of living
material per
area decreaces

Man	5th consumer
Trout	4th consumer
Frog	3rd consumer
Ladybird	2nd consumer
Aphid	1st consumer
Grass	Producer

Fig. 2.1.4   A pyramid of numbers

How the producers produce – photosynthesis

By definition, plants are able to produce their own food by the process of **photosynthesis**. The name, 'photosynthesis', really tells you that it is a process which uses the energy from light (photo) to make things (synthesis). All the classes of food (see p. 108) except minerals can be made by plants during this process.

Plants are able to take carbon dioxide from the air, water from the soil, and using light for energy, in the presence of green chlorophyll, can make food.

$$\text{Carbon dioxide} + \text{water} \quad \xrightarrow[\text{chlorophyll}]{\text{light}} \quad \text{Food} + \text{oxygen}$$

The process can be divided into two basic stages:

1 The splitting of water molecules using the energy from sunlight.
2 The adding of the hydrogen, from the split water molecules, to carbon dioxide molecules to form carbohydrates.

In order to produce proteins and some vitamins it is essential for plants to obtain a supply of nitrogen. Very few organisms can make use of the abundant nitrogen in the air and plants obtain their nitrogen from the soil as nitrates. However, if all plants took nitrates from the soil, the supply would soon be used up, so there is a cycle which ensures that the nitrogen used by plants is put back into the soil. This is called the **Nitrogen cycle** and it can be summarized as shown in Fig. 2.1.5.

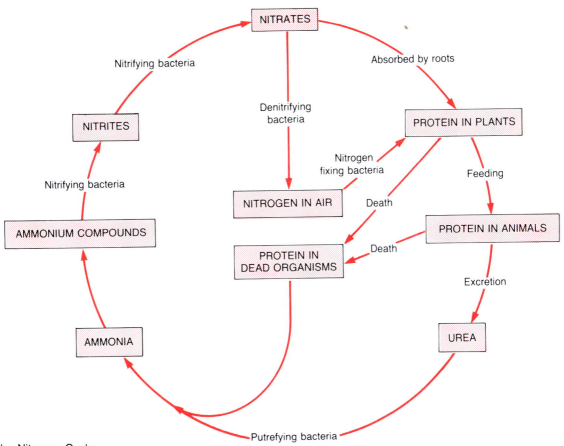

Fig. 2.1.5 The Nitrogen Cycle

This complicated looking cycle can be considered as being made by the building up of parts and the breaking down of parts, each using different types of bacteria. It is an excellent example of how useful bacteria can be to Man. In fact, there are many types of bacteria involved which are essential to all other living organisms.

Bacteria which build up things in this cycle are:

1 **Nitrifying bacteria**. These build up molecules of nitrates from nitrites and ammonium salts.
2 **Nitrogen-fixing bacteria**. These are most unusual because they are some of the very few organisms which can take the nitrogen from the air and change it into a form which can be used by plants to make proteins. They are found in the soil but also occur inside the roots of some special plants called **leguminous** plants, e.g. the pod-bearing plants such as peas, beans, clover, vetch and lupins.

Bacteria which break things down in the cycle are:

1 **Denitrifying bacteria**. These break down nitrates into nitrogen and oxygen so that they get back into the air.
2 **Putrefying bacteria**. These are responsible for the decay of dead things. They break them down to carbon dioxide and ammonia. Ammonia is such a reactive chemical that it forms ammoniun salts very quickly in the soil.

There is a similar cycle that ensures a constant supply of carbon dioxide to plants. It relies on the fact that the carbon dioxide they use for photosynthesis is put back into the atmosphere by animals and bacteria. This is called the **Carbon cycle** and it is shown in Fig. 2.1.6.

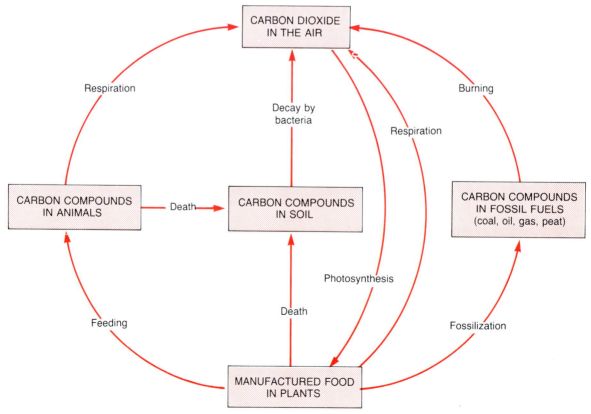

Fig. 2.1.6   The Carbon Cycle

Questions: Domains I and II

1 The diagram shows a simplified food web for an island.

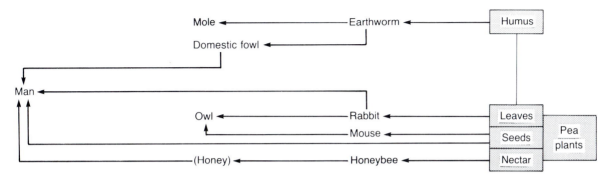

a Name the source of energy which makes the food web possible. (I)

b The mice were eliminated from the island because they became pests. This could affect the rabbit population in two ways. Explain both possibilities. (II)

c Use the food web and write down a food chain which has Man as the third consumer. (II)

d State precisely what Man should eat to give the least energy loss between himself and the producer in this food web. (II)

2 The apparatus was set up as shown. The colour of the indicator was noted in each tube at ten minute intervals. The indicator changes from cherry red to purple when it becomes less acid.

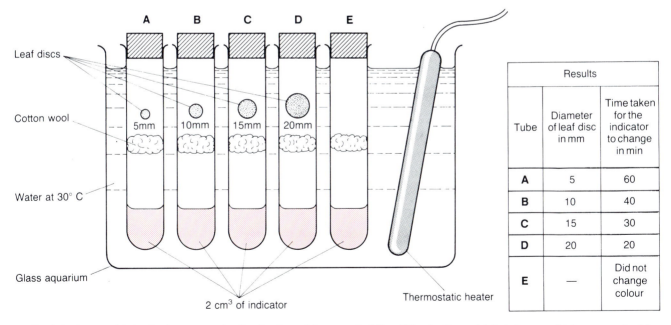

Results		
Tube	Diameter of leaf disc in mm	Time taken for the indicator to change in min
A	5	60
B	10	40
C	15	30
D	20	20
E	—	Did not change colour

a Explain how a biological process in the leaf could have caused the indicator to change from red to purple. (II)

b Why did the indicator in tube D change colour more rapidly than that in tube A? (II)

c State the purpose of tube E. (II)

d After 60 minutes which tube would be expected to contain most oxygen? (II)

e If the apparatus was kept in the dark for several hours, which tube would contain the most oxygen at the end of this period. (I)

Experimental skills and investigations: Domain III

1 Is light necessary for photosynthesis?

Starch is one of the substances produced as a result of photosynthesis. It is one of the easiest substances to test for in a laboratory because it goes from white to dark blue when iodine (in potassium iodide) is added to it. Using this knowledge, we can test whether photosynthesis has taken place in a leaf, by testing the leaf for starch.

Method
A suitable potted plant, such as a *Pelargonium* (geranium) is destarched by leaving it in a dark cupboard for two days. This is done to ensure that any starch present in the plant is changed to sugar. Chemical reactions take place in the plant, in the dark, which break down any stored starch into sugar. After this treatment we can be sure that no more starch is present in the plant.

If part of a leaf is now surrounded by aluminium foil, as is shown in the diagram, then we are now ready to investigate whether light is needed for photosynthesis.

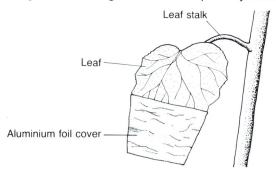

Leaf stalk

Leaf

Aluminium foil cover

The potted plant, with one of its leaves partly covered with aluminium foil, is left in bright sunlight for a day. The leaf with the aluminium foil is removed and tested for the presence of starch as follows:

1 It is boiled in water to break down the cells so that some of the green chlorophyll is released.
2 It is then heated over a water bath in ethanol to remove all of the chlorophyll. NEVER HEAT ETHANOL DIRECTLY BECAUSE IT IS HIGHLY INFLAMMABLE.
3 It is then soaked in water to soften the leaf and remove excess ethanol.
4 Iodine in potassium iodide is then added to the leaf and the observation noted.

What do you conclude from your observations?

2 Is carbon dioxide necessary for photosynthesis?

The starch test is used as a basis of this investigation as in the previous one.

Method
A destarched plant is set up as shown in the diagram and is left in bright sunlight for a day. The purpose of the concentrated sodium hydroxide is to remove carbon dioxide from the flask. After exposure to light, the leaf from the flask and the leaf in the normal air are both tested for the presence of starch in the way described previously.

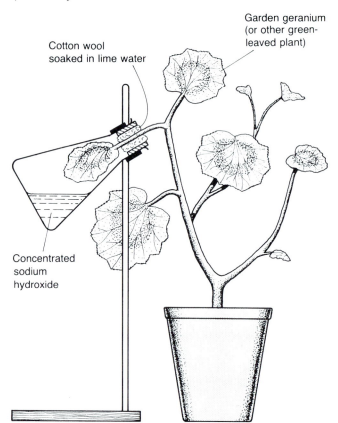

Garden geranium (or other green-leaved plant)

Cotton wool soaked in lime water

Concentrated sodium hydroxide

What do you conclude from your observations?

3 Is chlorophyll necessary for photosynthesis?

The starch test is used as a basis of this investigation as in the previous one.

For this test we can use a variegated leaf. This leaf has chlorophyll limited to parts of it with white parts in between. Many plants are variegated in this way. In the laboratory, suitable plants to use are *Pelargonium* or *Tradescantia*.

Method
A destarched plant with variegated leaves is left in bright sunlight for a day. After exposure to light, a variegated leaf is tested for starch in the way described previously.
What do you conclude from your observations?

2.2 Bugs, fleas and other nasties

(a) △

Man has survived in spite of the environment in which he lives. He uses its resources for feeding himself in competition with vast numbers of other organisms. An even greater number of organisms feed on Man (see Fig. 2.2.1). We call them **parasites**. An accurate definition of a parasite poses a problem. We could say that a parasite is a living organism that obtains its food from another living organism. If we use this definition, is a developing embryo inside its mother a parasite?

A parasite is a living organism that obtains its food from another living organism of a different species without giving anything in return. Some parasites of Man cause disease.

Why do we become ill?

There are many reasons. They might have nothing to do with parasites. The body may not function properly because its parts may be worn out or altered in some way. There could be a problem which was inherited from your parents. Perhaps you are not eating the correct type or amount of food. Stress may be a problem. Harmful substances from the environment might have entered your body. Some of the disorders caused by these factors are discussed in other sections of this book. Here we are concerned with diseases caused by organisms which live in or on your body.

If a parasite gets into human tissue, establishes itself and reproduces faster than our defence mechanism can cope with it, we call it a **pathogen**. Man has been able to survive the ravages of parasites by defending himself in many ways.

His first line of defence is his skin. It acts as a barrier. We only become aware of this when we cut or graze ourselves. If no treatment is given, the damaged area soon festers due to invading bacteria. We can produce our own chemical to kill bacteria (a **bacteriocide**). Tear glands, near our eyes make such a chemical. Hydrochloric acid is made in our stomach which kills bacteria entering on food.

Our blood system is also very important in defence (see p. 140). Blood clots seal wounds against bacteria. Special white bloods cells (**phagocytes**) eat microbes which may enter the body. Antibodies and antitoxins are made by some white blood cells (**lymphocytes**) which neutralize the effects of microbes.

Most microbes cause disease when they produce poisonous waste products. These poisons prevent normal chemical reactions from taking place in your cells.

Microbes reproduce very quickly. In the warmth of your body they may divide once every half hour. After a short time an enormous number of microbes may be causing problems. It takes time for your body's defence mechanism to work. During the early stages of an infection the pathogens multiply rapidly and you just begin to show symptoms of the disease. This is called the **incubation period**. When the symptoms are being shown to their full extent, this is called the **period of illness**. The **recovery period** is the time when the pathogens are being destroyed faster than they are reproducing. Pathogens are often carried from one person to another. If this happens then the disease is said to be **infectious**.

Highly infectious diseases must be reported to your doctor. Under some circumstances the doctor will inform the local Environmental Health Officer. This is true of 'notifiable diseases'. The aim is to trace the source of infection and reduce the spread of the disease. In Britain the

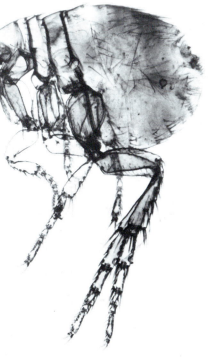

(b) △

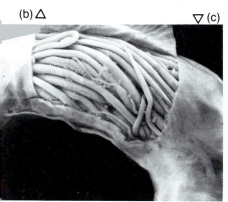

▽ (c)

Fig. 2.2.1 (a) A bed bug; (b) a flea and (c) nematode worms

following diseases are notifiable: measles, poliomyelitis, smallpox, yellow fever (caused by viruses); cholera, diphtheria, food poisoning, tetanus, tuberculosis, typhoid, whooping cough (caused by bacteria); malaria (caused by a single-celled animal).

Various countries have their own laws governing notifiable diseases.

Viruses and disease   Viruses are minute particles which can only reproduce within living cells (see Fig. 2.2.2). If they are outside living cells they cannot carry out all the characteristics of living organisms. Indeed, it could be argued that they are non-living.

Some of the more common viral diseases are discussed below.

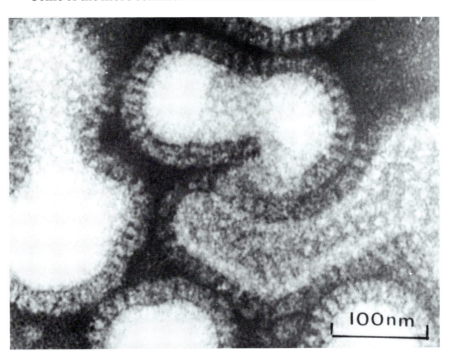

Fig. 2.2.2 Electron micrograph of a virus (influenza)

1 The common cold   The incubation period is three days. It is spread by air-borne droplets when you sneeze. During the period of illness there is irritation of the respiratory tract, coughing and sneezing. It is important to remain isolated for one to two days so that you will not spread the disease by coming in contact with other people. Aspirin is often given to relieve discomfort and, where necessary, to lower the abnormally high temperature. Influenza is similar but the symptoms are more severe. You should isolate yourself for seven days and secondary infections can result. These occur when the virus damages cells and bacteria can enter and feed on them.

2 Measles   The incubation period is ten days and the virus is spread by air-borne droplets. During the period of illness you develop a sore throat, a running nose, a cough and fever. A rash appears on your scalp, neck and ears. Small white Koplik's spots appear on the inside of the mouth. A similar disease is rubella (German measles) with an incubation period of 18 days. This is particularly serious in women during the first four months of pregnancy. The virus can be passed on to the developing embryo with a 20 per cent chance of the baby being born blind and deaf. Fortunately there is a method of prevention by vaccination (see p. 46). Girls between the ages of 11 and 14 have the opportunity of this treatment while still as school.

3 Mumps   The incubation period is 14 – 21 days. The virus is spread by saliva and air-borne droplets. Again, the disease can be prevented by vaccination. During the period of illness the patient suffers from high fever, swelling of the salivary glands, and the reproductive organs may be affected.

4 Chickenpox   The incubation period is 14 days. The virus is spread by direct contact because of droplets from open sores. Sores develop after a rash. There is a fever during the period of illness.

5 AIDS   **Acquired immune deficiency syndrome** (**AIDS**) is a disease which interferes with the body's normal immunity to diseases. It is caused by a virus and transmitted via the blood.

Since the first cases of AIDS were reported, much of the natural history of the disease has become clear. AIDS is the name given to a group of clinical conditions which includes several life-threatening infections, particularly one type of pneumonia. It was first observed among previously healthy homosexual men in the USA in 1981.

It is now clear that a cause of all AIDS is infection with a single, newly discovered virus – the AIDS virus. Homosexuality, promiscuity and drug abuse cannot cause AIDS without infection with the AIDS virus. Infection with the virus is potentially lethal to all men, women and children irrespective of lifestyle or sexual activity.

A few weeks after infection, the virus commonly causes an illness of short duration similar to glandular fever or influenza. This is followed by a period when no symptoms are shown which may last months or years. Later there is weight loss, fever, diarrhoea and infection by a large variety of bacteria, e.g. tuberculosis or pneumonia. Infection with the AIDS virus can be fatal even without any visible symptoms.

The AIDS virus is in a group of viruses called Lentivirinae, of which only three other species are known. The study of this group of viruses has largely been neglected because they appeared not to affect humans and they could not be transmitted to small laboratory animals. In domestic animals, lentivirus infections have proved so lethal and unresponsive to treatment, that slaughter of infected animals has been the universal means of control.

The AIDS virus infects white blood cells in the lymph nodes and in the spleen. It also infects cells throughout the brain. Any antiviral agent that only prevents multiplication of the virus must be continued for life. Any agent which destroys all cells containing the virus would also destroy brain cells. The AIDS virus kills those white cells (lymphocytes) which help the body to become immune to various diseases for a period of months or even years. The almost unlimited varieties of lentiviruses, combined with the inability of antibodies to eliminate the virus, has made ineffective all attempts to produce a vaccine to combat it. Protection against infection with the AIDS virus using existing vaccination techniques will take a long time to develop.

6 Poliomyelitis   This is a particularly serious disease because there is no known cure or successful treatment. Its incubation period is ten days. It is spread by air-borne droplets, water and food. Fortunately it can be prevented by vaccination. The worst symptoms include paralysis and wasting of muscles. This is a very dangerous problem when the respiratory muscles are affected. Respiratory movements have then to be performed by an 'iron lung'.

Bacteria and disease

Most bacteria do not cause disease, Indeed, many are essential for the life of plants and, indirectly, animals. The essential ones live in the soil or in the roots of certain types of plants. Some of these use nitrogen from the air and change it to nitrates which are taken into plants and used to make proteins. Others are essential for the processes of decay, in which many elements are recycled. There are occasions when Man uses bacteria in the manufacture of his food (see p. 114). However, the bacteria that usually have most publicity are those which cause disease. These are parasitic. They enter cells, pour out digestive enzymes and digest the cells. Their excretory products do most damage because they are poisonous. Reproduction is by simple division and is very rapid if conditions are suitable.

Probably the most common bacterial diseases in Britain today are those that cause food poisoning and other diseases of the intestine. These are transmitted by faecal contamination. In simple terms, this is due to lack of hygiene after using the toilet. If you do not wash your hands after using the toilet it is possible for egested waste (faeces) from your intestine to reach food. If the faeces carries bacteria then these will be able to enter the body during feeding. Diseases caused by such bacteria include typhoid, paratyphoid, dysentery, cholera and *Salmonella* food poisoning.

Fig. 2.2.3   Photomicrograph of bacteria (*Streptococci*, cause sore throats)

Some bacterial diseases

Food poisoning

About 75 per cent of all cases of food poisoning in Britain are due to *Salmonella*. Infections may be from food eaten in restaurants but home cooking may be responsible for many of the cases. Every year we read newspaper reports of outbreaks of food poisoning. Some of these have led to many deaths. Symptoms of severe abdominal pain and diarrhoea, roughly 24 hours after a meal often indicate *Salmonella* poisoning. Undercooked meat is a rich breeding ground for the bacteria involved. For example, in October 1977 more than 1000 school children in Stockholm developed *Salmonella* food poisoning after eating ham salad served in school cafeterias (see Fig. 2.2.4). The food had been prepared in a central kitchen and distributed to several different schools.

Fig. 2.2.4   The spread of *Salmonella* food poisoning

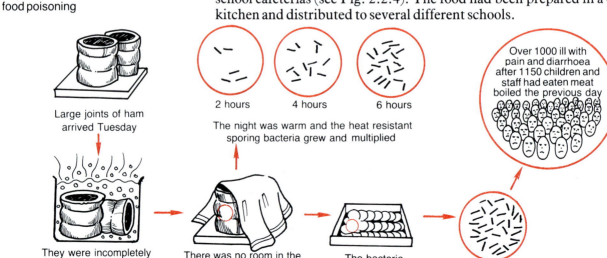

Large joints of ham arrived Tuesday

They were incompletely boiled Tuesday afternoon

There was no room in the refrigerator so they were allowed to cool slowly overnight

The night was warm and the heat resistant sporing bacteria grew and multiplied

2 hours   4 hours   6 hours

The bacteria multiplied

Over 1000 ill with pain and diarrhoea after 1150 children and staff had eaten meat boiled the previous day

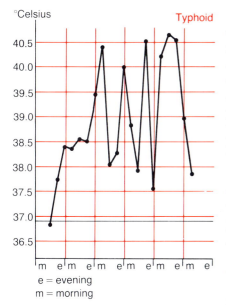

°Celsius   Typhoid

e = evening
m = morning

Fig. 2.2.5 Periodic fever caused by *Salmonella typhi*

Fig. 2.2.6 Paths in the spread of *Salmonella* food poisoning

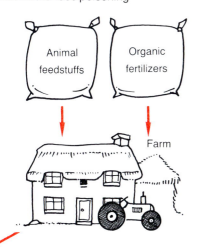

A particularly serious form of food poisoning is typhoid fever. It is caused by *Salmonella typhi*. The bacterium spreads into the blood stream from the intestine and affects many organs. The incubation period is about ten days during which time the bacteria spread via the lymphatic channels (see p. 155). High fever (see Fig. 2.2.5), headaches, chills, abdominal pains and constipation occur during the beginning of the period of illness. By the end of the first week severe diarrhoea occurs. In the second week of illness a rash appears and the liver and spleen enlarge. By the third week the patient is weakened by fluid loss. The intestine may become perforated and death may result.

Before modern medical care and antibiotic treatment, about 12 per cent of all patients died of exhaustion. Today, the risk of death is less than two per cent. Antibiotics (see p. 47) such as ampicillin and chloramphenicol have made recovery almost certain if the disease is recognized in its early stages. Vaccination against typhoid was first carried out in 1897. The vaccine called TAB (typhoid and parathyphoid A and B) is commonly given to travellers to Mediterranean and tropical countries.

It is possible for some people to carry the bacteria inside them and yet not suffer from the disease. Perhaps the most well known of all such typhoid 'carriers' was Typhoid Mary. She was a domestic cook in America. She managed to infect 57 people, resulting in three deaths. The bacteria spread from one household to another during her employment. Eventually, she was detected as a carrier. It was rumoured that she was cooking nearby when an epidemic of 1300 cases occurred in New York in 1903.

The spread of *Salmonella* food poisoning depends on live bacteria in food (see Fig. 2.2.6). Another type is caused by *Staphylococcus*. This type only requires the presence of the poison produced by the bacteria. The poison is not broken down by heat. It remains active even if the bacteria are killed by boiling. In 1975 during a flight from Tokyo to Paris, 195 out of 364 people on board a 'Jumbo Jet' became ill. The cause was ham which had been reheated on the plane. The ham had been contaminated by a carrier of *Staphylococcus*. It had not been stored at a low enough temperature. The bacteria multipied and produced the poison. Reheating the ham increased the multiplication. Fortunately, none of the flight crew was ill because they had not eaten the ham. After this incident the World Health Organization recommended that flight crews should receive meals from two different sources. The chance of all the crew being affected during one flight would thus be very small.

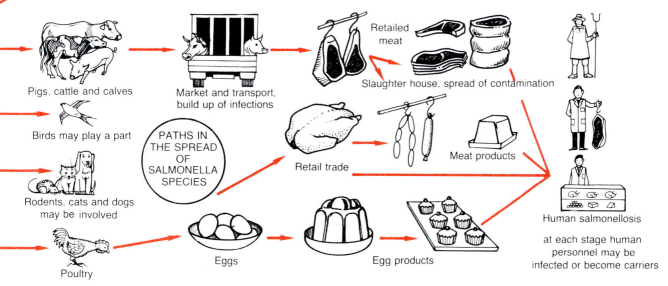

Pigs, cattle and calves

Market and transport, build up of infections

Retailed meat

Slaughter house, spread of contamination

Birds may play a part

PATHS IN THE SPREAD OF SALMONELLA SPECIES

Retail trade

Meat products

Rodents, cats and dogs may be involved

Human salmonellosis

Poultry

Eggs

Egg products

at each stage human personnel may be infected or become carriers

Tuberculosis   **Tuberculosis** is one of the many diseases spread in microscopic droplets of air-borne saliva or on dust particles. Diphtheria and pneumonia are spread in similar ways. *Mycobacterium tuberculosis* is the bacterium which causes tuberculosis – still a major cause of death in some parts of the world.

The lungs are most commonly affected but any tissues can be infected as the bacteria spread. Symptoms include massive loss of weight, coughing, fever and tiredness. The person may cough up blood and became anaemic (see p. 145). Normal tissue breaks down and is replaced by tubercles or nodules.

Success in the treatment of tuberculosis depends on rest, good ventilation, and a nourishing diet. Antibiotics are the most effective drugs. The vaccine used for immunization against tuberculosis is made from weakened (attenuated) live cultures of the strain of bacterium which normally occurs in cattle. It was first used in 1921. Since 1951, mass vaccination programmes have been carried out throughout the world. Consequently it is very uncommon in Britain today.

Fungi   Among the body's parasites is a large group of organisms called **Fungi**. They are plant-like in form and have branching bodies. However, they do not have chlorophyll and so cannot make their own food. Their feeding method is similar to that of bacteria. Food is digested outside their bodies by digestive enzymes that are poured from the tips of their branches. Some types invade our bodies, especially the respiratory system. Others are confined to our skin. The most common are ringworm and athlete's foot. Thrush is a common fungus which grows inside the mouth or in the vagina.

 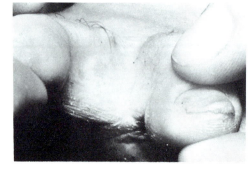

Fig. 2.2.7 Ringworm infection   **Fig. 2.2.8** Athlete's foot infection

Ringworm (Tinea)   This fungus feeds on the outer layers of skin (see Fig. 2.2.7). It is very infectious, being carried on skin fragments, combs, scissors, towels and pets. Fungi reproduce using spores. These are tiny parts which behave like the seeds of flowering plants, although they are much simpler and formed in a different way. Spores are resistant to heat and drought. They can be blown around in the air or attached to insects. Therefore they are easily spread. **Ringworm** forms circular patches on the skin and is very unsightly when it spreads over the body. Because it is so infectious, children suffering from it must stay away from school until it is cured. **Athlete's foot** is similar (see Fig. 2.2.8). It is caused by *Tinea pedis*. It gets its name from the fact that it is easily spread via changing rooms and swimming pool areas. The fungus digests the skin between the toes.

Treatment is by the fungicide, nistatin, and zinc creams applied to the skin. There is an antifungal antibiotic called griseofulvin which is taken in tablet form. It is absorbed into the blood stream and taken to the skin. Here it protects the outer layers against invasion.

Animals made of one cell – Protozoa

A parasitic single-celled animal causes more deaths in the world than any other disease. It is called *Plasmodium* and it causes **malaria**. Malaria is a disease that occurs mainly in the hot, damp, tropical areas of the world (see Fig. 2.2.9). Certain types of mosquitoes are responsible for carrying the parasite from one person to another.

Fig. 2.2.10 Photomicrograph of *Plasmodium* (malaria parasites) in the blood

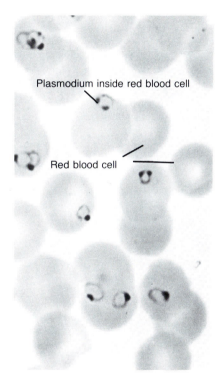

Plasmodium inside red blood cell

Red blood cell

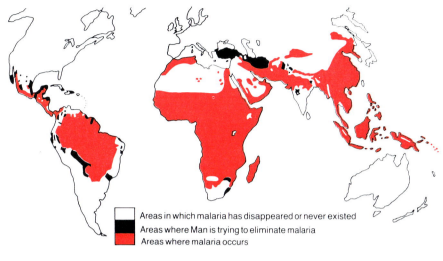

Areas in which malaria has disappeared or never existed
Areas where Man is trying to eliminate malaria
Areas where malaria occurs

Fig. 2.2.9 Map showing distribution of malaria

Most mosquitoes are harmless but in the tropics some types called *Anopheles* infect Man with malaria. Female *Anopheles* feed on the blood of vertebrates so they can obtain a protein to help them make their egg cases. When an infected mosquito bites a person, the tiny *Plasmodium* passes into the victim's blood (see Fig. 2.2.10). The mosquito has needle-like mouth parts for piercing skin. Once in the blood, the parasite goes through a complex life-cycle. It enters the liver and multiplies rapidly by dividing repeatedly. Then it enters red blood cells and continues to reproduce by dividing. The newly formed parasites burst out of the red cells. It is at this stage that poisonous waste products are released, causing fever (see Fig. 2.2.11).

Some of the parasites change into sexual stages which act like sex cells. If these are taken back into a mosquito when it feeds, they will join together. As a result of this sexual reproduction, a zygote is formed. It is in the form of a cyst which is lodged in the mosquito's stomach wall. Many young stages burst out of the cyst. They move to the mosquito's salivary glands and infect new hosts when they are bitten. The cycle then begins all over again.

Until the 1940s there were about three million deaths per year from malaria. The health authorities began a campaign to control the disease via the World Health Organization. Drugs such as quinine and chloroquine kill the parasites when they are inside Man. Paludrin and daraprim can be used to prevent infection if they are taken regularly in malarial areas.

Attempts have been made to destroy mosquitoes with insecticides and by draining swamps where they lay their eggs. Problems have arisen because mosquitoes have become resistant to many types of insecticides. Some types of *Plasmodium* have developed resistance to some anti-malarial drugs. Spraying breeding areas with oil will kill the larval stages of the mosquito. However, there are many cases where this is not practicable. One example is in paddy fields where water must be present for rice to grow. In these areas, fish have been introduced to eat the larvae in the water. Recently, progress has been made on developing a vaccine

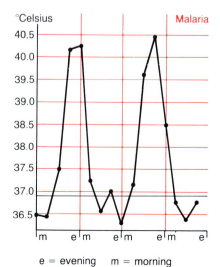

°Celsius   Malaria

e = evening   m = morning

Fig. 2.2.11 Fever caused by *Plasmodium*

against the parasite. However, use of such a method on a world-wide scale will probably be too expensive in the countries of the Third World where the greatest problem of Malaria remains. Most success in eliminating malaria has been on islands. For example, Singapore and Cyprus are now free from the disease whereas, previously, it was a problem in these countries.

Entamoeba histolytica
An unpleasant gift
from the tropics

Dysentery is a term that is often used wrongly to mean any form of diarrhoea. While there are numerous causes of diarrhoea, including infections of almost any part of the alimentary tract, very few of these diseases should be described as dysentery. There are two common types of dysentery – bacillary, caused by a bacterium and amoebic, caused by a single-celled animal. The main feature of the disorder is that an infection causes the inner lining of the large intestine to become ulcerated. This leads to the loss of blood, which together with body fluids and dead cells passes into the gut. This mixture is rapidly expelled from the body as blood-stained and liquid faeces.

Amoebic dysentery is caused by the single-celled animal *Entamoeba histolytica* (see Fig. 2.2.12). It normally lives as harmless parasite on the inner surface of the large intestine. However, under certain conditions, the parasite invades the gut wall and causes ulcers to form. From the intestine the amoebae may invade the portal veins that lead from the gut to the liver. When this happens the amoebae may multiply in the liver to form an abscess. Patients with amoebic dysentery pass large numbers of amoebae in their faeces. These are not infectious to other people. They soon die even if swallowed.

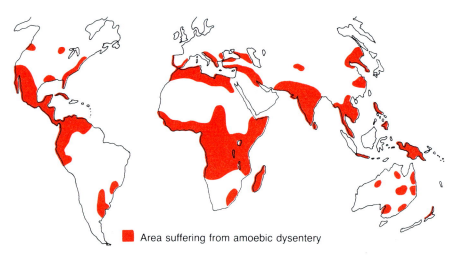

■ Area suffering from amoebic dysentery

Fig. 2.2.12   Map showing distribution of amoebic dysentery

Infection occurs when the amoebae form cysts and these are swallowed. These cysts, because of their thick walls, will survive for up to two months in suitable conditions. They are produced by amoebae living harmlessly on the surface of the large intestinal wall. Many amoebic infections cause no symptoms at all and the host is unaware of the infection. At any time, however, tissue invasion may occur and cause dysentery and liver abscesses. These illnesses, especially liver abscesses, can occur many years after the patient was originally infected. A carrier may remain infectious for 25 years.

An amoebic infection becomes established in the large intestine within three or four days after the cysts have been swallowed. Symptoms may be delayed for several weeks or even years. The ulcers are often deep beneath the wall of the gut.

Cure and prevention

Treatment can be divided into two stages: (1) that given to kill the parasite and (2) measures used to treat the patient's general condition. It is often the latter that can save the life of a seriously ill patient. Loss of body fluid must be replaced, either by mouth or in severe cases via a vein. If bleeding has been extensive, a blood transfusion will be necessary.

Injections of emetine are usually necessary for several days, followed by another drug to eliminate successfully the infection from the gut. Metronidazole, given by mouth, is used with considerable success. The treatment of liver abscesses involves draining them of pus with a needle. This is followed by emetine injections together with the anti-malarial drug, chloroquine.

Until all communities have a high standard of hygiene, Man will continue to be attacked with dysentery. The infected food handler, the housefly and direct contamination of water supplies or food, all transmit the parasite. Food handlers are very important in the transmission of *Entamoeba* because it is often the carrier with no symptoms who infects others.

Worms

Bilharzia (Schistosomasis)

About 250 million people in the tropics are infected with **bilharzia** (see Fig. 2.2.13). It is a great handicap in the struggle of many people of the developing countries to achieve a reasonable standard of living. Bilharzia is caused by a flat worm or fluke (*Schistosoma*). It lives as a parasite inside the body, feeding on the carbohydrate in the blood. Eggs of the worm have been found in mummies of Eygptian Pharoahs. It was not until 1852 that Bilharz, a German doctor, discovered that the disease which now bears his name was caused by a parasitic worm. In 1907 the species of worms causing the disease was identified.

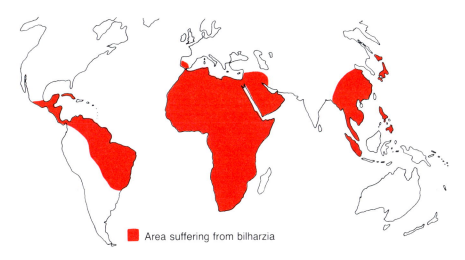

■ Area suffering from bilharzia

Fig. 2.2.13 Map showing distribution of bilharzia

How do you suffer from bilharzia?

People get bilharzia by swimming or wading in fresh water containing the young stages (larvae) of the fluke. These tiny larvae are torpedo shaped and fast moving with a forked tail. When they come in contact with human skin they attach themselves, drop their tails and then bore through the skin. They are powerful enough to penetrate clothing. Bilharzia infection can also be caused by drinking water which contains the larvae.

In this case the larvae bore through the lining of the mouth and throat. Once inside the body the worms swim in the blood, growing in size as they feed. They can grow up to 22mm long and have a life span of 30 years. Male and female worms occur. After the male and female mate the female lays large numbers of eggs for many years.

The effect on the victim

Bilharzia does not kill directly but it does have terrible effects. The most serious is anaemia. The victim loses blood because the worms are actually feeding on his blood and because of the damage caused by the eggs. The eggs have sharp spikes at one end. They pass into the intestine and urinary bladder. They tear the soft surfaces of these organs causing internal bleeding. Eggs and blood eventually leave the body in the urine and faeces. Victims are being bled to exhaustion day after day.

How is it spread?

The victims have millions of eggs produced inside them by the worms. When they urinate or defaecate the eggs leave the body. If they reach fresh water they hatch and become tailless larvae. These larvae have to find a host within 48 hours or die. The only suitable host is a type of fresh water snail. In fact, bilharzia is sometimes called 'snail fever'. If the larvae find a snail of the right type (*Bulinus* or *Planorbis*), they penetrate it and breed. In about four weeks they emerge as mobile tailed larvae. These swim around and must find a human host within a day or two if they are to survive. The odds against an egg hatching into a larva – finding the right kind of snail in time – hatching into a tailed larva and then finding a human host, are enormous. However, each worm produces so many millions of eggs that although a vast number die, sufficient survive to spread the disease. Often those with bilharzia will urinate and defaecate into ponds, streams and canals. People may use this water for swimming, drinking and washing. A cycle is set up in which affected people continually infect and reinfect themselves and others.

Treatment

Drugs can be injected into veins or muscles to cure bilharzia. The treatment may take a long time because it is so easy for a person to become reinfected. Prevention is the best cure and is only achieved by better hygiene. It is important to advise people not use water suspected of carrying bilharzia larvae. Boiling drinking water makes it pure. If water for washing is allowed to stand in containers for four days, larvae will be dead by the time it is used. Latrines should be made well away from rivers. This would prevent victims from pouring millions of eggs into water and thus keeping the cycle going. Building bridges over streams so that people would not have to wade through them would also help.

Attempts have been made to destroy the snails, the other host of the parasite. If the snails could be destroyed then the cycle would be broken and the disease gradually eliminated. Copper sulphate has been used. The fruit of the endod shrub also kills the snails. The fruit of this shrub contains natural soaps. It is used in some areas of Ethiopia for washing clothes. The problem with using chemical snail killers (molluscicides) is that they are often harmful to Man and useful animals. There is also the problem of treating large volumes of moving water frequently. Snails breed very rapidly.

The future
Bilharzia is one of the few diseases affecting large numbers of people that appears to be increasing. This has been caused by the spread of irrigation. This increases food production and helps to prevent malnutrition but has the effect of extending bilharzia. Irrigation canals are the ideal breeding grounds for the intermediate host snails.

Recently, progress has been made in the development of a completely different method of control. This involves running the victim's blood through a filter machine, filtering out the parasites and then returning the blood to the body. It is almost like a kidney machine (see p. 175). While effective, this treatment requires skilled operators and complex equipment. It is unlikely that it could be used as mass treatment in the developing countries of the world. Here the answer must be improved hygiene.

Roundworms

Roundworms belong to the group **Nematoda**, many of which are important parasites of man. The larvae (young stages) often cause most of the symptoms of infection. The adults often live in the intestines. They mate and the eggs of larvae leave the body in the faeces. There are two ways in which the larvae may get back into a human body. The eggs are picked up on contaminated fingers, or they are carried to food by insects and are swallowed and hatch in the intestine. This occurs in *Ascaris lumbricoides* (Fig. 2.2.14), the life-cycle of which is shown in Fig. 2.2.15.

Fig. 2.2.14 *Ascaris lumbricoides*

The life-history of Ascaris lumbricoides (see Fig. 2.2.15).
The egg is swallowed in food and drink (stages 1 and 2). The cyst covering the egg dissolves and the larvae escape (stage 3). They burrow through the wall of the small intestine (stage 4) and enter the blood stream, eventually reaching the lungs (stage 5) by way of the liver and the heart. These larvae are then carried up the trachea (stage 6) and to the throat. They are swallowed and once again pass through the stomach to the small intestine where they become adults. The eggs of the new adults are egested in the faeces (stage 7). Some are unfertilized (stage 8) and some are fertilized (stage 9). Many of the fertilized eggs never develop. In favourable conditions, however, the infective larvae develop (stage 10) and are once again capable of contaminating food.

The hookworms have free-living larvae, which live for a while in soil contaminated by faeces before they enter through the skin of people walking barefoot. In these cases, it is the adult worm in the intestine that causes the symptoms. With *Ascaris*, the presence of a number of worms may cause diarrhoea, pain in the abdomen, or even blockage of the intestine. The hookworms live on blood and tissue fluids, thus causing anaemia, blood loss and nutritional deficiency.

Enterobius vermicularis, the **pinworm**, deposits its eggs near the anus. They contaminate bedclothes and the fingers. The infestation causes severe itching and the infected person, often a child, cannot resist scratching the area. The eggs on the fingers then get to the mouth and are swallowed to cause reinfection or a new infection.

Intestinal worms are killed or eliminated by treatment with drugs called anthelmintics.

All the diseases transmitted via faeces can be prevented by proper hygiene. Faeces should be disposed of where flies, food or bare feet cannot be contaminated.

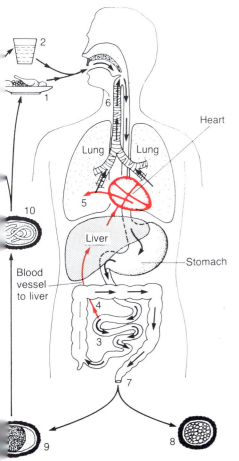

Fig. 2.2.15 The life-cycle of *Ascaris lumbricoides*

Hookworm

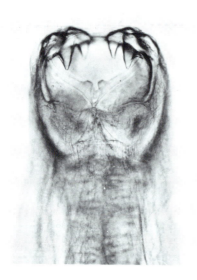

Fig. 2.2.16 The head of a hookworm – the vampire that drains blood from more than 900 million people

The disease resulting from the parasitic hookworm has plagued mankind for a very long time. There is a reference to this disease in an Egyptian papyrus dated 1500 BC. However, these parasites were not fully recognized until 1843. They are parasites of Man and of other mammals in the tropics. They live in the small intestine, gaining entrance by burrowing through the skin. They are roundworms or nematodes. Two types infect Man, *Ancylostoma* and *Necator*. The adults live on the wall of the intestine, attached by their horny teeth (see Fig. 2.2.16). Here they feed on the mucous membrane and blood. A heavy infection can cause severe anaemia. Females lay as many as 25 000 eggs each day. They pass out of the host in the faeces. Larvae hatch in a day or two and feed on bacteria and organic debris in the soil. Each larva moults four times. They stay up to 15 weeks in the soil. Then, if they make contact with human skin they bore their way into the body (see Fig. 2.2.17). They are carried to the heart by the blood stream and then go to the lungs. As soon as they reach the respiratory organs they are carried up the air passages by cilia. Finally they are coughed-up and swallowed. They reach the small intestine and attach themselves to the wall.

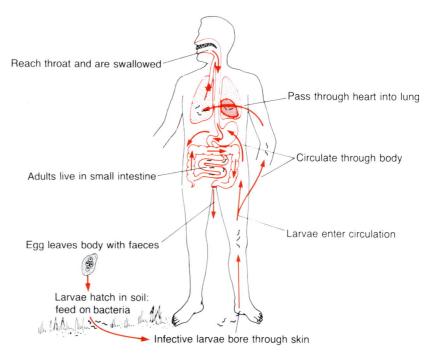

Reach throat and are swallowed

Pass through heart into lung

Circulate through body

Adults live in small intestine

Larvae enter circulation

Egg leaves body with faeces

Larvae hatch in soil: feed on bacteria

Infective larvae bore through skin

Fig. 2.2.17 The life-cycle of the hookworm

The growth stages and the journey through the body has taken five weeks. Most of the worms are eliminated during the next year. Some remain in the intestine for as long as 16 years. As a disease agent, hookworm can devastate a whole population. Large infections cause serious anaemia and diarrhoea. Adult worms can be expelled from the intestine with medicines called vermifuges. Successful control depends on an understanding of the life-cycle. The use of proper latrines and the wearing of shoes have dramatically reduced the incidence of hookworm in many areas of the world.

Lice and fleas

Lice are small, flat, wingless insects which are parasites on the skin of mammals and birds. They live on skin, hair and feathers. Those that suck blood will feed on mammals, including Man. Three species live on Man: *Pediculus humanus* – variety *capitis* on his head, variety *corporis* on his body, and *Phthirus pubis* in the pubic hair (see Fig. 2.2.18).

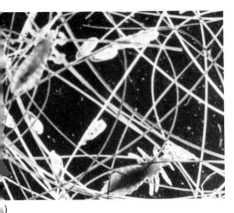

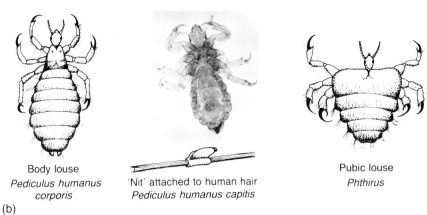

Body louse
Pediculus humanus corporis

'Nit' attached to human hair
Pediculus humanus capitis

Pubic louse
Phthirus

(b)

Fig. 2.2.18 (a) Lice and eggs on human hair and (b) the types of lice living on humans

The lice live in the hair or clothing, often in seams, and do not move about much outside their chosen area of the body. Each female crab louse (*P. pubis*) produces up to 50 eggs during her life time, while the body louse can produce 300. The eggs (nits) are attached to hairs or clothing. Lice live from two weeks to a month, depending on temperature, humidity and food. They may last several days between feeds.

When feeding, the louse injects saliva which irritates the skin, producing a raised itching red spot. Scratching may cause infection from bacteria living under the finger nails.

The body louse, and to a lesser extent the head louse, carries typhus and trench fever. Because it does not like high temperatures, it moves to another person as soon as its host becomes ill and feverish, thus spreading the infection further.

Lice infestation is discovered by finding their nits in people's hair. Treatment in crowded insanitary living conditions is difficult. Insecticides, repellents and thorough washing of the body, hair, and clothes with anti-louse preparations will clear up an infestation. Insecticides that are effective against lice include benzene hexachloride, DDT, benzyl benzoate and benzocaine. These can be used as powders or combined in a lotion. Repeated applications are usually necessary.

Fleas are members of a group of insects which has no wings. They have piercing and sucking mouth parts and large hind legs for jumping (see Fig. 2.2.19). In the adult form they are all parasites of mammals and birds.

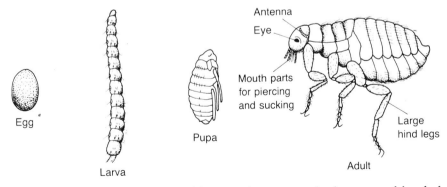

Egg

Larva

Pupa

Antenna

Eye

Mouth parts for piercing and sucking

Large hind legs

Adult

Fig. 2.2.19 Stages in the development of a flea

Fleas exist all over the world among humans and other warm-blooded animals. The most common species attacking Man in the United Kingdom are the human flea, *Pulex irritans*, the cat flea, and the dog flea. They are small brown insects, 1.6 mm long, which can hide in clothing and jump very actively if disturbed. They bite humans around the waist and legs and are usually most troublesome in homes with cats and dogs. Rat fleas and tropical sand fleas also bite Man.

Fleas bite through the skin, extract their food from the blood and cause itchy red patches. If a person has many bites, prolonged scratching may cause the bites to be infected. The irritation is caused by the injection of a fluid produced by the insect's salivary glands. Some people are very sensitive to this fluid and react very badly to flea bites.

In addition to skin problems which fleas cause, they can also spread serious diseases. The rat flea transmits plague and typhus.

The best way to control fleas is with insecticides sprayed on carpets, floors, and furniture. Household pets should be examined as a source of the infestation and de-flead with an insecticide. Their blankets and kennels probably harbour fleas and should be treated accordingly.

For treatment of human bites, a soothing lotion containing menthol, camphor, or phenol can be applied to the skin. Special powders can be prescribed which contain an insecticide for use on the body and on clothing to kill the fleas. Antihistamine creams should not be used because so many people are sensitive to them.

Control of disease

Immunity

The body tends to reject harmful organisms that enter it. The ability to do this is called **immunity**. We are constantly coming in contact with harmful viruses and bacteria. If we did not have our immune responses, we should always be suffering from diseases and ill-health. Certain cells, **lymphocytes** (see p. 156) produce **antibodies** when they come into contact with invading organisms. Protein molecules on the outside of the parasites trigger the lymphocytes' responses. The antibodies neutralize the effects of the invading parasites.

Throughout your life you are naturally immune to certain diseases which occur in other animals and vice versa. For example, your pet dog cannot catch measles. Immunity to a particular disease may be inherited. Often people are not affected by chicken pox virus even if they come in contact with it. There are ways of making you artificially immune. This is **acquired immunity** and it is induced by **vaccination**. A short-term 'passive' protection may be obtained by using **serum**. Serum is plasma without the ability to clot. An animal is given a mild form of the disease. It reacts by producing antibodies. If an extract of the animal's serum is purified and injected into Man, the antibodies retain their effect. This type of treatment is only given in an emergency, for example, an anti-tetanus injection. If a person is suffering from diphtheria, this method is also used.

In the eighteenth century, the English doctor Edward Jenner (see Fig. 2.2.20) discovered that people could be protected against the dangerous disease, smallpox. He did this by injecting them with the related, but quite harmless, cowpox virus. Since Jenner's day a mass world-wide vaccination programme has completely eliminated smallpox. Today, this type of 'active' acquired immunity gives long-term protection. There are three main methods. They all depend on the fact that lymphocytes produce antibodies as a result of contact with the surface proteins of the parasite. This happens whether the parasite is living or dead – the lymphocyte merely reacts to the shape of the surface molecules.

The first method relies on injecting a vaccine containing a weakened or attenuated virus; for example, smallpox or poliomyelitis. The second method is by injecting a vaccine containing dead virus; for example, whooping cough (see Fig. 2.2.21). The third method involves injecting a modified poison, produced by the harmful micro-organism. This makes the body produce anti-toxins which give protection against an attack.

Fig. 2.2.20   Edward Jenner

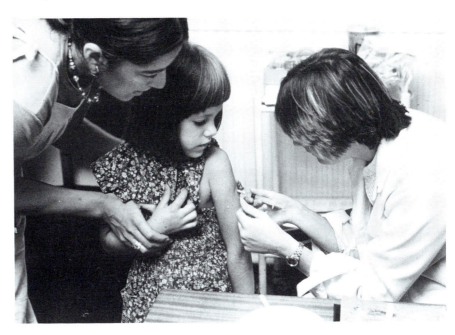

Fig. 2.2.21 Modern vaccination – a child being vaccinated

Antibiotics

When you are infected with bacteria, the doctor often gives you an **antibiotic**. This is a substance produced by a living organism that stops some living things from growing. **Penicillin** is the one we usually think of first but there are many others. New ones are constantly being discovered. Before antibiotics were discovered there were few ways of curing bacterial diseases. Antibiotics cure many of them so fast that they were called 'miracle drugs'.

The green or grey 'fuzz' that grows on old bread, fruit and cheese in the kitchen is the colony of a fungus. Some fungi have developed a special defence. They make a poison that prevents others from living too close to them. This poison is called an antibiotic because it also prevents bacterial growth.

Until the nineteenth century, no one knew what caused infectious diseases. Then, in 1860, the French scientist Louis Pasteur showed that some diseases are caused by **bacteria**. By the 1890s the bacteria that caused many diseases and infections were known. Doctors began to apply bacteria-killing liquids to wounds. However, these did not prevent internal infections.

In 1928, Alexander Fleming discovered penicillin. He was studying *Staphylococcus aureus*, a bacterium that causes may infections. He grew that bacteria in Petri dishes. The bacteria grew and multiplied on top of the culture medium in the dish until it was covered with a golden, creamy layer. One day, Fleming found that a few cells of mould had contaminated the dish. Around each piece of fungus was a clear area where bacteria were not growing (see Fig. 2.2.22).

The mould was producing an antibiotic. He grew more mould and collected the droplets of liquid that it made. He tested the liquid and found it stopped bacterial growth. The mould was *Penicillium notatum*. Fleming named his new drug penicillin.

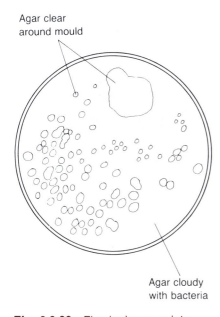

Agar clear around mould

Agar cloudy with bacteria

Fig. 2.2.22 Fleming's agar plate

Biotechnology and antibiotics

In 1939, two scientists in England, Dr Howard Florey and Dr Ernst Chain, tried to produce large amounts of penicillin. The problem they faced was that, as the fungus grows, it uses up all its food and air. It produces waste products that collect and harm it. Biotechnologists helped the doctors to solve these problems. They designed methods to keep bringing new food and air and to take away waste products. By 1943,

Fig. 2.2.23   Mass production of antibiotics

penicillin could be produced in 1000 gallon tanks (see Fig. 2.2.23). Then a new species of *Penicillium* was found that makes 200 times more penicillin than Fleming's species does.

Penicillin cures many diseases, but not all. The search for antibiotics continues. Some of these are still produced by growing colonies of micro-organisms. Others are now produced by man-made processes. Antibiotics work in various ways. Some kill bacteria. Others stop them from multiplying. One way they do this is by interfering with the making of the coats surrounding bacteria. Then the body is able to kill the bacteria.

A major problem in using antibiotics is that bacteria become **resistant** to them. Then they are useless. There are two types of resistance. Resistance to penicillin builds up if small doses are given. It can be avoided by giving such large doses of penicillin that all the bacteria are killed at once. But resistance to streptomycin can occur all at once. It is not predictable. People who take antibiotics for a long time may develop new infections. Our bodies are full of micro-organisms. They all compete for food. If an antibiotic kills the strongest, the others have an opportunity to grow. Often the new infection is caused by a fungus. Fungi are often unaffected by antibiotics that harm bacteria.

Antiseptics   **Antiseptics** are chemicals that kill bacteria, or stop them multiplying. They are used to get rid of bacteria from the skin, clothing and furniture. Antiseptics help to prevent diseases from spreading. Generally, only bacteria are killed by antiseptics. If the skin is scratched or cut, bacteria infect the wound. The body can fight back, but needs some help if there are too many. The cells inside our bodies are delicate and can be harmed by strong chemicals, so antiseptics are used only on the outside of the body. Really strong antiseptics, that would even harm the skin, are called disinfectants. They are used, for example, on table tops and floors.

Lister in 1865 first used antiseptic techniques during surgery. He invented a carbolic acid spray which created an antiseptic atmosphere around the operating table. Some types of bacteria produce hard cases around themselves, called spores. Antiseptics cannot kill these. An antiseptic also does not sterilize. Sterilization is the killing of absolutely all bacteria. It is normally done by boiling surgical instruments and other utensils. Antiseptics reduce the number of bacteria present.

Soap is a weak antiseptic. Surgeons scrub their nails, hands and arms thoroughly before operating. There must be as few bacteria as possible in the operating room. Bacteria in an operation can cause disease and prevent proper healing. Alcohol, in a 70 per cent solution, is one of the antiseptics used to clean your skin before an injection.

Biotechnology combats disease:
interferon

Interferons were discovered in 1957. They are hormone-like proteins that are part of the body's defence against viruses and other disease-causing agents. The possibility of extracting and purifying interferons, and using them to boost the production of natural interferon during illness has excited researchers for many years. Yet in the 1970s human white blood cells were the only source of interferons, and the world's supply amounted to no more than a few milligrams. Recently scientists have produced strains of bacteria that can produce vast amounts of interferons. Some strains of bacteria now make more than a milligram of interferon per litre of culture (see Fig. 2.2.24).

Fig. 2.2.24   Production of interferon

More than 2000 cancer patients have now been treated with one type of interferon and about one-sixth of these patients have shown a positive response. Obviously then, interferons are not a miracle cure for cancer.

There are as many types of cancer as there are types of cell in the human body. Different cancers respond in different ways to interferons. Eighty per cent of patients with a certain type of leukaemia had treatment which made all the affected cells disappear from the blood and reduced the amount of cancer cells in the bone marrow. In this type of leukaemia, interferon may be the best treatment. In general, cancers of the white cells of the blood and bone marrow respond to interferon, whereas the more common cancers of the breast, intestine, and lungs do not.

The side-effects of the use of interferon present a problem. Patients have experienced many reactions. All patients have mild to moderate fever, tiredness and muscle pain. They feel as though they have influenza. As doses get higher the patients suffer hallucination and coma.

Three main types of interferon – alpha, beta, and gamma are produced by different cells. The alpha interferon can be produced commercially by whole human cells infected with a virus, as is done in large vats by biotechnology.

Interferons act against viruses. They act directly on cells, not the viruses that infect them, and there is evidence to suggest that interferons limit the course of virus infection naturally in animals and people. Interferon will not cure the common cold but may prevent it. Alpha interferon is also useful in other viral diseases such as the liver disease hepatitis, and various illnesses caused by herpes viruses, such as shingles.

Interferons offer us a new approach to the treatment of disease. They are the first of a whole series of agents that will become available. These substances are produced naturally by our bodies as part of our defence against infectious agents. In using them, we are trying to help reactions that already exist.

Questions: Domains I and II

1 A pure culture of non-pathogenic bacteria was grown in nutrient broth. 1cm^3 was taken and placed in a sterile Petri dish to which nutrient agar was added and thoroughly mixed. After the agar had set, three wells of equal size were cut with a cork borer. The wells were half filled with different toothpastes and then the dish was incubated for two days. The results are shown in the diagram.

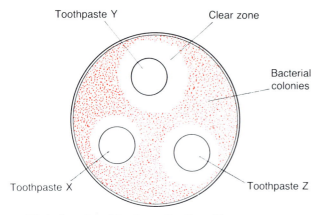

a State the aim of the investigation. (I)
b What type of substance might the toothpaste contain to have the effects shown? (I)
c What is meant by 'non-pathogenic'? (I)
d How could you sterilize the cork borer? (I)
e At what temperature should the Petri dish have been incubated? (I)
f Which toothpaste was most effective against this type of bacterium? (II)

2 A mystery disease caused by an unknown microbe affected a town. People suffering from the disease were coughing, suffered pains in the chest and a rash appeared on their faces. There were some people who did not contract the disease although they had been in contact with sufferers. The graph shows the body temperature of a sufferer after infection.

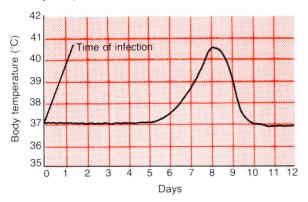

a What is the incubation period of the disease? (II)
b On which day was the temperature at its highest? (II)
c What was the highest temperature reached? (II)

d What was probably happening in the microbes between
  i days 5–8?
  ii days 8–10? (II)
e What is the normal body temperature? (II)
f Apart from the rise in body temperature, name two other symptoms of the disease. (II)
g Some of the people were naturally immune to the disease and some acquired immunity when the same disease struck months later.
  i What is meant by natural immunity?
  ii What is meant by acquired immunity? (I)
h How many diseases are notifiable diseases. Who has to be notified of such a disease? (I)
i Name one notifiable disease. (I)
j Diseases may be spread when the microbes causing them are breathed into a person's lungs. Write down three other ways in which diseases can be spread and give one example of each. (I)

3 One of the major factors restricting the economic development of Africa is the incidence of the disease **sleeping sickness**.

Sleeping sickness is a disease endemic in Africa. It is caused by a protozoan, *Trypanosoma*, which lives as a parasite in the plasma of blood and affects both the nervous and lymphatic systems.

Trypanosomes are transmitted from one person to another by tsetse flies. After being bitten by the vector, the lymph glands swell and later the patient becomes apathetic and sleepy, suffering muscular spasms and uncoordinated movements. Eventually coma and death follow. Drugs used so far are very toxic.

The following passage is taken from *The Guardian* of 27 January 1982.

Sleeping sickness, the potentially fatal disease that affects about 35 million people in Africa, might soon be controlled by drugs which attack the fly-borne parasite which causes it.

Work at the Medical Research Council's National Institute for Medical Research at Mill Hill in London, now suggests a range of existing drugs might be modified to make them active against the organism.

It has been found that an anti-tumour antibiotic called Daunorubicin is, in the laboratory, several thousand times more effective against any parasite than any existing drug.

Initial hopes were dashed when the drug was tested in animals, because it simply did not work. Dr J. Williamson, of the Division of Parasitology at Mill Hill, explained yesterday that the failure arose because, in the blood stream, as distinct from the laboratory test systems, the drug passed through the parasite without having time to be effective. What was needed was some way to make it stick.

Earlier research had shown that many substances, including serum albumin, are taken into the parasite cell, so an attempt was made to marry the drug to serum albumin to see whether this would serve as a carrier and specific delivery system.

The first results from trials on experimental animals have confirmed that it works exceptionally well. But the fact that a method of targeting has been identified could mean that older drugs, such as the poisonous arsenical compounds, abandoned because of their dangers, could be brought back into use.

Explain the meaning of the following:
a (i) endemic; (ii) protozoan; (iii) parasite; (iv) vector; (v) antibiotic. (I)
b Why do you think the lymph glands swell in a patient suffering from the disease sleeping sickness? (I)
c Why do you think the patient should suffer muscular spasms and uncoordinated movements? (I)
d Why was the antibiotic not effective when animal experiments were carried out? (II)
e How was this problem solved? (II)
f What do you understand by 'a method of targeting'? (II)
g What types of toxic drug were formerly used? (II)
h How is it that older toxic drugs might now be used again? (II)

4 The graph shows two methods by which immunity to a disease may be brought about.

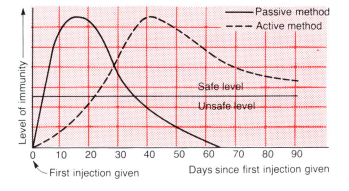

First injection given
Days since first injection given

a Which method loses its effect quickly? (II)
b How long after injection with the active method does it take for immunity to reach a peak? (II)
c For how many days does the passive method provide a safe level of immunity? (II)
d Which method do you think would be most suitable for someone who had just spent a day with an infected person? (I)
e What is the difference in immunity between the active and passive methods 15 days after injection? (II)

5 Here is a list of stages which led to an outbreak of typhoid:

1 River water was passed into a water purification works.
2 The purification treatment failed.
3 The water left the water works and was used in a factory producing corned beef.
4 The corned beef was exported to Britain.
5 Cans of corned beef reached the shops.
6 A meat slicer was used to cut the corned beef as well as other cooked meats.
7 Two weeks later people suffered from typhoid.

a From where did the typhoid microbes first come? (I)
b Explain how the corned beef became infected. (I)
c Explain how it could happen that some customers at the shop, who only bought ham or bacon, also became ill with typhoid. (I)
d What name is given to the period of time between the customers eating infected corned beef and becoming ill with typhoid? (I)
e Give any two measures which are taken by local health authorities to prevent outbreaks of typhoid in the community. (I)
f The events in the list could lead to an epidemic of typhoid. Name three conditions which would cause an outbreak of a disease to be classed as an epidemic. (I)

Experimental skills and investigations: Domain III

How does a fungus feed?

1 Prepare a Petri dish with a starch/agar mixture.
2 Place a smal piece of *Mucor* (about 1mm^2) in the centre of the Petri dish and incubate the Petri dish at 25°C for two days.
3 Flood the Petri dish with iodine solution. Pour off the surplus iodine and hold up the Petri dish to the light.
4 Record your observations.

Questions
a What evidence is there to suggest that the starch has been broken down?
b What evidence is there to suggest that the starch is not broken down in one step?
c The way in which ringworm feeds on human skin is similar to the feeding method of *Mucor*. From the evidence of your observations suggest how feeding takes place.
d How would you test to see whether the starch had been digested to a reducing sugar?

2.3 Nice to be near?

Your teeth

Your teeth should be yours for life. If you take care of them, they will last as long as you do. Neglect them and you will suffer the pain and distress of toothache. You will find it difficult to eat and speak properly. Healthy teeth and gums are an essential part of an attractive smile.

There are two main types of dental disease:

1 **Periodontal disease** – which affects the gum, the bone and the tissues surrounding the tooth and
2 **Dental caries** – tooth decay.

Both diseases are caused by plaque.

What is plaque?

Plaque is a sticky film made of bacteria and material from saliva (see Fig. 2.3.1). It is almost invisible and forms on the teeth, around the gum margins, and between the teeth. If it is allowed to remain bacteria irritate the edges of the gums causing inflammation – the first sign of periodontal disease.

If plaque is not removed it combines with chemicals in the saliva which make it become hard. This cannot be removed by brushing. More plaque clings to this hard deposit called **calculus**. More bacteria breed, increasing the inflammation of the gums.

As well as causing periodontal disease some of the bacteria in plaque change sugar into acid. This may dissolve the tooth enamel and start the process of dental decay.

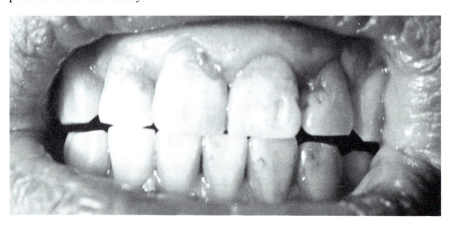

Fig. 2.3.1 Plaque

Plaque can be stained with disclosing tablets.
These are chewed and then swished around the mouth and over the teeth to stain the plaque red or blue. By staining plaque in this way you can tell whether you have cleaned your teeth properly. Having seen where most plaque appears you can now pay special attention to these areas.

Periodontal disease

An enormous number of undecayed teeth are extracted every year because of periodontal disease. If often starts in young teenagers and is a frequent cause of bad breath.

Periodontal disease affects the area surrounding the tooth and is caused by an accumulation of plaque. Calculus may form making it even more difficult to remove the plaque effectively.

The first stage of periodontal disease is **gingivitis** which is an inflammation of the gums. It is very common among teenagers and is caused by bacteria in the plaque. The first signs are redness and swelling of the edge of the gums which often bleed when brushed. It is usually painless so that many people ignore the problem. At this stage gingivitis can usually be cured by thorough tooth cleaning to remove the plaque.

If teeth are not cleaned properly and gingivitis continues the inflammation may spread around and down into the root of the tooth, destroying the fibres which hold the tooth in place (see p. 126). A pocket forms between the tooth and the gum where more bacteria can collect. It is then almost impossible to remove all the plaque and the disease gets worse. Eventually the teeth become loose and may have to be removed.

How to brush your teeth It is more important to clean everywhere thoroughly than to worry about a particular method but the technique described here works for most people (see Fig. 2.3.2).

1 Place the tufts of the brush pointing to the junction of the teeth and gums at an angle of about 45 degrees to the teeth.
2 Vibrate the brush in a circular motion gently round the necks of the teeth and between them.
3 Brush the backs of the front teeth.
4 Brush the biting surfaces with a backwards and forwards scrubbing movement.

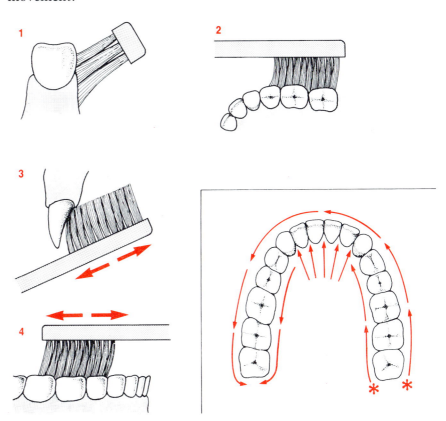

Fig. 2.3.2 How to brush your teeth

It is important to keep to a routine so that no area of the teeth remains unbrushed. Begin at the back of the lower jaw and work around the outer surfaces. Then do the same with the inner surfaces. Repeat the process for the upper jaw. Then brush the biting surfaces of both upper and lower teeth. From time to time you should check that you are removing all the plaque with a disclosing tablet.

Make sure your tooth brush has medium tufts so that they won't damage the teeth or gums. Replace it regularly.

Problems with your skin

Acne – a teenage worry

Fig. 2.3.3 Acne

Acne is one of the commonest problems to affect a teenager's skin. It may show as just a few blackheads on the chin or as masses of large purple spots covering the face, neck, shoulders and back (see Fig. 2.3.3). Severe forms of acne occur in only three to five per cent of people and must be treated by a doctor. The mild forms occur in nearly everybody, usually during teenage years. Acne is not a sign of uncleanliness but this can make it worse. You should realize that after a year or two, for most people, the acne will disappear.

What is acne?
Puberty is caused by chemical messengers called hormones. An imbalance in the body's hormones often leads to acne.

The sebaceous glands of the skin (see p. 177) produce oil. This normally keeps the skin supple but when there is too much of it the skin becomes greasy. The oil prevents the outside layers of skin from being shed. It allows the pores of the sebaceous glands to become blocked. The glands continue to produce oil. This cannot escape and stays in the entrance to the gland mixing with dead skin cells and bacteria to form a hard plug. If the pore is large then the plug is exposed to the air. Chemical changes turn the plug to a black colour, resulting in a blackhead. Its colour is not caused by dirt but by the chemicals of the mixture. If the oil remains under the skin because the pore is very small, then it remains white and is called a whitehead.

Blackheads look unsightly but because the oil can leak away they do not usually become infected with bacteria. Whiteheads cause oil to build up in the passages of the glands. This build up may become infected as the bacteria feed on the oil and cells. White blood cells move to the area to destroy the bacteria. Dead white cells and bacteria form the yellow pus. This becomes the familiar spot or pimple.

In severe cases of acne the walls of the sebaceous glands break, spilling their contents on to the surrounding tissue. This leads to a much larger spot which becomes swollen and purple as more white blood cells move in to combat the infection. Squeezing spots often results in this because the infected mixture is forced on to otherwise healthy skin.

Controlling acne
You cannot stop acne, but you can limit its effects. The first and most important rule is to wash thoroughly. A wash in the morning is essential, and showering or bathing after exercise is also important. Washing with a mild soap or cleanser will help remove the oil and layer of dead cells. This prevents blockage of the sebaceous glands. Scrubbing the skin or the use of soaps with powerful grease removing chemicals may remove all the useful oil. The skin then dries, flakes and looks unsightly. Flannels should be boiled frequently and lather removed with warm water before drying. Some of the blackheads may be removed with brisk rubbing with a towel.

Certain soaps contain antiseptics which may destroy bacteria living on the surface of the skin. However, if the bacteria are deep within the sebaceous glands this type of soap may be of little use.

Emotional tension has been shown to affect acne. Spots may subside during school holidays only to appear again when term starts. Pre-menstrual tension in girls may make the problem worse, just before a period starts. It may also become worse during exams or if you have a job interview. There is little you can do about this except keep to the basic rules of hygiene. If blackheads are removed by squeezing then the following rules apply:

1 First wash your hands and thoroughly scrub your finger nails.
2 Press some cotton wool on to the blackhead.
3 Gently squeeze the area to remove the blackhead entirely.
4 Dab the area with a little antiseptic.

Spots, especially the large purple blotches, should never be picked. There is too great a risk that the pus will be squeezed out through the side of the sebaceous gland and on to the skin. This increases the infected area. The after effects of picking spots may be much worse than the spots themselves. Wounds caused by finger nails will leave the skin permanently scarred and unsightly, especially on the face.

There are many creams and lotions available from chemists for use on acne. They are unlikely to have much effect on many of your spots because spots begin deep within the skin. The creams are not able to reach far into the skin and therefore cannot really get at the infection. The acne creams will have some effect on blackheads. They work by peeling away the blackheads and unblocking the pores. They contain resorcinol, salicylic acid or benzoyl peroxide. Any non-greasy make-up used to hide spots must be thoroughly removed each night.

Occupations involving working near oil are known to encourage acne.

Hair care The best way to keep hair healthy is to keep it clean. A good shampoo should be used with gentle massaging of the scalp. You should massage the whole scalp to stimulate the blood vessels and remove dead skin. Carefully rinse off the shampoo using clean warm water until your hair 'squeaks' (see Fig. 2.3.4).

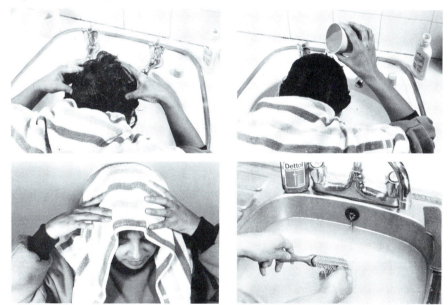

Fig. 2.3.4 Keeping your hair clean

When drying your hair, pat it dry with a towel but do not rub too hard because this could lead to hair breaking. When using a hair-dryer avoid using very hot air. This tends to affect the natural oils of the hair and can make the hair brittle.

Regular brushing helps keep the hair in good condition because it removes dirt and stimulates the blood flow to the skin. Combs and brushes should never be shared because this can lead to the spread of parasites between people. Combs and brushes should also be cleaned in a little disinfectant.

It is not true that cutting the hair will make it grow more thickly. If the ends of the hair are found to be splitting, they simply need to be cut. Split hairs just mean that the layers of hair have separated at the ends. Singeing is not advisable as this may dry the hair.

Lacquers and mousses, widely used to hold hair styles in place, are made mainly from gum in alcohol and water. They are safe to use for normal hair, but make it lose its natural sheen and lustre if used in excess. Most popular brands of permanent waves (perms), tints and dyes are also safe to use on healthy hair, providing over heating is avoided and they are not used repeatedly. Too frequent perming or straightening of the hair may cause splitting, as may constant bleaching. Repeated pulling at the hair may also damage it, and so it is unwise to wear hair tied back tightly all the time, or to roll the hair in curlers every night. Such harsh treatment of the hair may lead to hair loss. Setting the hair in rollers, fairly loosely, and for short periods should not prove harmful in this way.

Dealing with dandruff   One of the most common scalp complaints is **dandruff**, otherwise known as **dry seborrhoea**. The skin of the scalp, like that of the rest of the body is constantly being shed in layers and scales. If there are a lot of waste products around on the scalp, scales of dandruff may result. It usually occurs just at the front of the head but sometimes over the whole scalp. If untreated, it may develop into a condition where flaking is more obvious and scales are yellowish in colour. Dandruff does not cause complete loss of hair if left untreated. It is still not known for certain what causes the condition. It may be bacteria, the result of malnutrition, a symptom of over-tiredness, or a disorder of normal scalp lubrication. Getting rid of the symptoms is often a problem but there are antiseptic shampoos which can control it. Only through such thorough and regular cleansing and care can dandruff sufferers expect their hair to remain healthy.

Questions: Domains I and II

1 The graph opposite shows the effect of eating sugar on the pH level of the mouth.
a What effect does the sugar have on the pH of the mouth? (I)
b What causes this effect? (I)
c How long after eating sugar does it take for the pH to reach a level where decay occurs? (II)
d For what length of time does the pH remain in a range where decay occurs? (II)
e What is plaque? (I)
f State three precautions which can be taken to reduce or prevent tooth decay. (I)
g Name two substances which must be present in a diet for healthy tooth formation. (I)

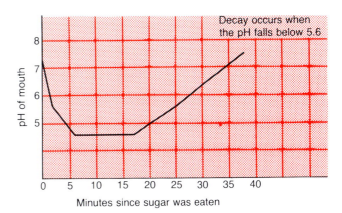

Minutes since sugar was eaten

2 Two towns, A and B, were selected from a survey to find the effects of fluoride on tooth decay. 250 children from each town were monitored for several years and the results are shown in the graphs. Only one town had fluoride added to its drinking water.

Town A

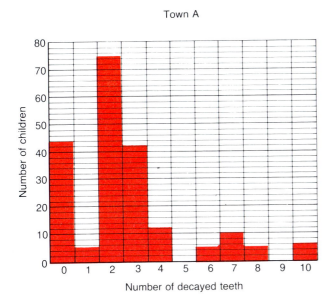

Number of decayed teeth

Town B

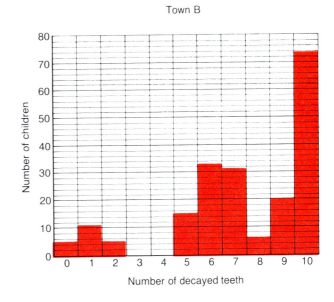

Number of decayed teeth

a Which town had fluoride in the drinking water? (II)
b Give a reason for your answer. (II)
c How many children in town A had only
  i two decayed teeth?
  ii ten decayed teeth? (II)
d How many children had three or more teeth decayed
  i in town A?
  ii in town B? (II)
c What percentage of children in town A had seven decayed teeth? (II)
f Name three other factors which could affect tooth decay. (I)

3 Some pupils investigated the amount of heat retained by three fabrics used for clothing. They covered one boiling tube with wool, one with nylon, and one with cotton. The same amount of hot water was put into each tube and a stopper carrying a thermometer was placed in each one. They took the temperature of the water every five minutes for 20 minutes. Their results are shown below.

Time (in minutes)	Temperature (°C) of water in boiling tube covered by		
	Wool	Nylon	Cotton
0	84	82	81
5	82	72	74
10	76	61	60
15	72	53	48
20	70	50	42

a Which fabric allowed the most heat loss in 20 minutes? (II)

b Which fabric allowed the most heat loss in the first 5 minutes? (II)

c Which of the three fabrics would you choose to wear on a cold day? (II)

d Give a reason for your answer to c. (II)

e If different amounts of water had been put into each tube at the start of the investigation, how would this effect the results? (I)

f What external condition might affect the results? (I)

g What else should the pupils have done to make their conclusion valid? (I)

The following extra information concerning the fabrics was obtained:

	Absorbancy	Washability	Strength
Wool	Quite good	Shrinks in hot water	Strong when closely woven
Nylon	Very poor	Dries rapidly washes easily	Very strong
Cotton	Very good	Can be boiled does not shrink	Quite strong

h Give two reasons why:
  i nylon clothes are useful on a summer camping holiday;
  ii cotton is used to make underwear;
  iii underwear should be washed regularly. (II)

i As well as clothing, our homes need to be clean. Explain the importance of:
  i Using a lid on a dust bin.
  ii Using disinfectant in a toilet.
  iii Having a bend in water pipes leading from homes into drains.
  iv Placing left over cooked meat in a refrigerator. (I)

Experimental skills and investigations: Domain III

How effective is your shampoo?

Procedure
1 Take four clean Petri dishes and four clean microscope slides. Label one of each A, B, C and D.
2 Take 25 cm³ of four different shampoos and pour each into a separate Petri dish.
3 Paint one side of each glass slide with a thin layer of cooking oil, stained with red eosin dye.
4 Start the stop clock and quickly put the slides, oil side down, into the Petri dishes. Rest the slide on the edge of the dish as in the diagram. Place slide A in dish A, slide B in dish B, and so on.
5 After five minutes remove the slides and lay them oil side up, on a paper towel. Compare the slides to see which shampoo has removed most of the oil.

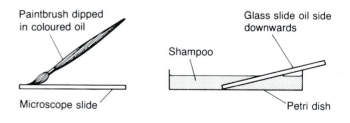

6 Replace the slides in the Petri dishes for a further five minutes. Remove them again and compare your result to see which shampoo is most effective at dissolving the oil.

Devise a similar experiment to compare the effectiveness of various soaps in removing oil.

2.4 Safe as houses?

A home full of hazards

Few people are aware of how dangerously most of us live our everyday lives. Roads, offices, schools, homes, and even our gardens, contain many hazards which could cause injury.

The majority of modern homes contain potential electrical hazards. Worn or overloaded circuits, or absence of proper earthing are common. Injuries result from dim lighting, uneven floors, irregular steps, slippery surfaces, obstacles, steep inclines, poor drainage, faulty ventilation, to name but a few. Broken glass and rusty cans litter playing areas. Deaths in Britain exceed 7000 each year from home accidents. In the USA the figure exceeds 30 000 people. The majority of all accidents in Britain occur in the home.

Most people find safety a fairly boring subject – until someone is crippled or killed for the lack of it. Of course, we automatically take many safety precautions every day. Our parents told us a thousand times to look each way before crossing the road, and so we do. But as a built-in defence against having too much to worry about, none of us can believe that any really bad accident will happen. So we keep forgetting to mend the broken stair someone might trip on, and don't bother fastening car seat-belts for short journeys.

Safety starts with intelligent design. Basically the designer must identify potential hazards and eliminate them. Poor design is responsible for the fact that colour television set owners have found smoke rising from their sets. One in forty report explosions. The problem is the high voltages needed for colour reception which heat up the sets. The heat should be carried away. Modern televisions are too small and tightly constructed to allow natural air circulation to serve as the coolant. Research is essential into heat resistant components. More is needed from television designers than good looking bits of furniture.

Whether at work, play, home, or in transit, design can only go so far to protect people. So much material of the modern world is potentially dangerous that people must be educated to be careful and alert.

Do-it-yourself is generally a dangerous way of saving money unless the home handyman becomes an expert (see Fig. 2.4.1). Few are! Britain's Consumer Association tested people on their knowledge of a revised colour code for wiring electric plugs. This was done on the day the code was introduced after months of publicity intended to educate the public. Only two-thirds tested said they knew the code. Of those, half had memorized it wrongly.

The chief reason for the very high number of home accidents is probably that people relax their guard in familiar surroundings. Long familiarity makes them unable to see even the worst hazards in their own homes. Faithful old washing machines, electric irons and favourite brands of cleansing liquid can all be a danger.

Home hazards can be divided into two classes. One involves unsafeness of home construction and major furnishings. The other is industry's shame – the vast number of unsafe goods that find their way into our homes (see Fig. 2.4.2). In the first case the householder should identify the hazards and arrange for their repair. Of course there are exceptions, such as when contractors have done poor work or when landlords fail to maintain property. The Consumer Association estimates that one out of six British homes has been dangerously left with unearthed electricity. In older homes electricty was often earthed along metal water

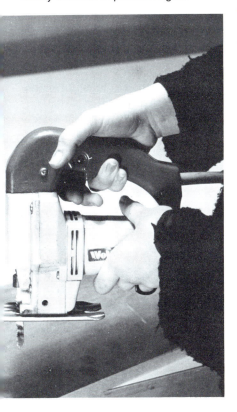

Fig. 2.4.1 Dangerous do-it-yourself equipment like this electric saw needs safety devices to protect fingers

Fig. 2.4.2 'Capacity 4 pints' – but it proved unsafe

Fig. 2.4.3 The Ronan Point disaster in London in 1968 when an explosion wrecked a block of flats

pipes running down the building. In many cases the ancient outdoor plumbing was replaced with modern plastics, without telling the householders they would need new earthing.

Electricity is a special problem in the bathroom. Wetness can turn a switch into an electric executioner. Never touch a switch or plug while standing in a bath, or even before drying off. Only carefully installed and placed heaters should be used in chilly bathrooms.

In older homes, particularly, overloaded circuits are a real fire hazard. People often plug in more and more labour-saving appliances. Too often people construct a network of extension wires instead of calling in an electrician to alter a circuit. It is hard to recognize inadequate wiring before it is too late.

Gas, a clean and efficient source of energy, may also be a highly dangerous one (see Fig. 2.4.3). Fortunately, gas, unlike electricity, smells. Gas Boards purposely add scent to odourless gases like the methane pumped out of Britain's North Sea gas wells. In the event of a suspected gas leak, the supply should be turned off and the leak promptly reported, not patched up. Never search for one with a naked flame.

Carbon monoxide is the odourless, colourless, killer component of gas. It is also produced when any fuel-burning appliance does not work efficiently. Carbon in fuel should be completely oxidized to harmless carbon dioxide. Faulty or ill-ventilated stoves, heaters and hot-water tanks burn fuel incompletely, producing carbon monoxide instead.

Fortunately, progress has eliminated many hazards from modern homes. Gas pipes and electric power lines are less likely to be exposed. Central heating cuts fire hazards by eliminating a large number of heaters and fires. Fuel is then burned in one place.

New hazards for old

It may be that we are introducing new dangers to replace the old ones. Widespread use of synthetics has made many modern living rooms ready to burn in a few minutes. Urethane foam-cushioning, chair frames, acrylic curtains and upholstery are all extremely flammable. In addition, fumes from certain smouldering plastics are poisonous. They can creep upstairs and kill after a family has gone to bed and need not burst into flames to do it.

Although the situation is improving, our houses are not generally designed for safety. Architects even more than other designers are influenced by people's expectations of what a house ought to look like. They seldom consider the best safety features. Electric sockets are usually set into the skirting boards. Yet Britain's Consumer Association recommends 39 inches off the ground as the best height for safety. This eliminates the need for the toe-touching bends that promote falls. It keeps trailing wires from lamps and appliances off the ground so that no one will trip over them.

More sockets in each room would similarly eliminate mazes of trailing wires. In old people's housing, which is specially built with high sockets, there are fewer reports of accidents from falls.

Researchers have found the kitchen sadly lacking in safe design (see Fig. 2.4.4). Poor design includes stove controls placed so that the cook must reach over boiling pots to switch on more heat. A great deal of kitchen equipment is too heavy and therefore easy to drop. Pots and pans with handles that become hot are obvious risks. Very high kitchen cupboards often find people perched on books, boxes and chairs. The washing machine is often the worst offender among home appliances. A poorly installed one may have the water supply dangerously near electricity.

Fig. 2.4.4 Researchers have found the kitchens sadly lacking in safe design

A tour of almost any house will reveal similar types of hazards in most rooms. The garden is also a potential danger area. In fact, with their many poisonous chemicals and sharp tools, gardens are particularly hazardous. Besides the obviously poisonous nature of insecticides and weed killers, garden chemicals can have surprising properties. The Consumer Association reports that entire lawns have caught fire after application of the weed killer, paraquat. It also found many power lawn mowers were inadequately guarded to prevent users from cutting off their toes. Further, power mowers cast aside small obstacles with the force of shell fragments. Electrically powered garden tools carry extra risks because of wet grass and the possibility of cutting through power cables.

Nor are rechargeable battery-operated tools always safe. Some, when plugged in for recharging allow current through to the metal blade. This could electrocute any curious child who might touch the blade.

Ladders for both indoor and outdoor use are too often shaky. It is always worth paying extra for a ladder that can be firmly anchored to a wall, or one that stands firmly on its own.

Two groups at risk — Old people and children are the groups most often involved in accidents. The presence of either in the house – even temporarily – requires careful thinking about hazards.

With a child around, almost anything seems dangerous! One would, however, think that at least toys would be safe. Unfortunately, this is not always so. There are at least four classes of dangerous toys: dolls and stuffed animals that might expose sharp parts when broken; darts (now only sold in sports shops); noisemaking toys with parts that could cut or puncture, and breakable rattles with parts that could be inhaled or swallowed. Designs are continually being tested and exposed as dangerous.

Curious toddlers like to stick pins and fingers into plug sockets and play with electrical appliances (see Fig. 2.4.5). Fortunately, child-proof socket covers are now on the market.

The aged represent a very different set of problems. Old people suffer terribly from falls. Their brittle bones break easily and heal very slowly. Overpolished or uneven floors, loose stair carpets, poorly lit stairs, all threaten the lives of the elderly. Falls are the greatest cause of injury in the home. The elderly are especially at risk in the bathroom. With decreasing strength and sense of balance, even raising oneself off the toilet seat can be a challenge. This is the reason why modern hospitals put sturdy grip bars at the side of each toilet and bath. In fact, every home should have bars for firm balance in the bath and shower. Even the fittest sportsman can slip on soapy porcelain. For the same reason, suction grip rubber bath mats should be used on the bottoms of baths.

Bathrooms are a further hazard, particularly to children, because dangerous chemicals are often kept there (see Fig. 2.4.6). The same hazard occurs in the kitchen. Most people tend to keep cleansers, bleaches and other poisons in under-the-sink cabinets within a child's reach. Manufacturers complicate problems with delicious looking labels. Who could not forgive a child for thinking that a bottle containing liquid with real lemon was not washing-up liquid but lemonade?

Keep all dangerous chemicals out of a child's reach and hide ladders. Pouring chemicals into edible drink bottles is asking for trouble, and not only from children.

Manufacturers rely heavily on attractive packaging to sell their products. Any law designed to make poisonous products look poisonous

Fig. 2.4.5 Toddlers like to play with electrical appliances

Fig. 2.4.6 Dangerous chemicals may be found in bathrooms

is therefore doomed to heavy resistance. A new protective law is always hard to have passed. It took Britain seven years to pass a law governing minimum safety standards for electric blankets even though many were known to have caused the deaths of users.

Hygiene – you and your home

The one place where eveyone likes to feel free from disease is at home. Even in the best kept household infection and allergies can be dangerous. There are some basic requirements which can reduce these dangers.

Strict hygienic measures are essential in the kitchen. All its working surfaces are breeding grounds for bacteria and fungi. Systematic cleaning of all traces of food after meals and scouring of draining boards, sink, cooker and all other working surfaces should be carried out. Floors, shelves and food storage places need the same attention regularly (see Fig. 2.4.7). All utensils need to be cleaned with a suitable scourer. Neither hot nor cold water alone can act as a powerful germicide so disinfectants should be used.

Fig. 2.4.7 Cleaning the kitchen

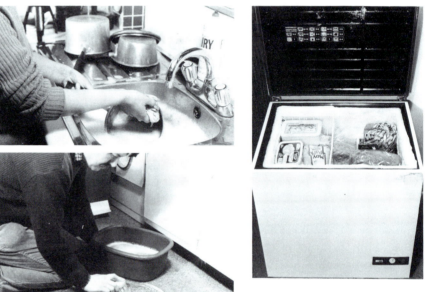

Fig. 2.4.8 A household freezer

Food storage is most important in household hygiene. People tend to do the bulk of their shopping on one day of the week, thus stockpiling foods. The most obvious hygienic treatment is to cover foods. This will not only protect against bacteria, but also prevent access by other household pests. An even greater protection is available in the form of the refrigerator (see Fig. 2.4.8).

Additional food should be stored in the larder of another cool room. As a general rule, all foods that are to be re-cooked must first be brought to a suitably low temperature, and then cooked at temperatures which will destroy bacteria.

Milk, a food which is quickly attacked by bacteria, can be safely stored in sterile containers or disposable plastic bottles. In either container it must be kept cool, particularly in the summer (see Fig. 2.4.9).

Preserved fruit and vegetables are best kept in special jars with air-tight seals. An air-tight seal is also important for drinks, which may be corked or kept in screw-top bottles. Before freezing, vegetables should be blanched by dipping them into boiling water. This kills bacteria and denatures any enzymes which might digest the plant cells.

Fig. 2.4.9 Some methods of storing food

Waste disposal Just as careful methods are needed to preserve foods, care is also needed in the disposal of waste food. Decaying material will always attract scavengers such as rats, mice, and insects when it is allowed to collect in the open or in drains. The most hygienic way of dealing with waste is regular cleaning of the sink with disinfectant or washing soda (see Fig. 2.4.10). Bins containing waste should always be tightly sealed in order to reduce contamination by pathogens (see Fig. 2.4.11). Disposable and sealable bags are the best way of getting rid of such waste.

Few bacteria flourish in dry dust in the home, but it is a health risk for many asthmatics. It may contain material which, when inhaled, causes an allergic response. Types of mites invariably infest house dust and often cause respiratory problems to develop when inhaled (see Fig. 2.4.12). Regular vacuum cleaning is essential to reduce this hazard, followed by the disposal of collected dust (see Fig. 2.4.13).

Pets are often considered a part of the family, but for the purposes of hygiene they need more than just affection. Both cats and dogs have to be cleaned and brushed regularly to remove dust and fleas that collect in their coats. Pets can carry organisms that cause disease in Man. Both cats and dogs can transmit ringworm, mites, viruses and worm infestations.

Fig. 2.4.10 Cleaning a sink   **Fig. 2.4.11** A tightly-sealed dustbin

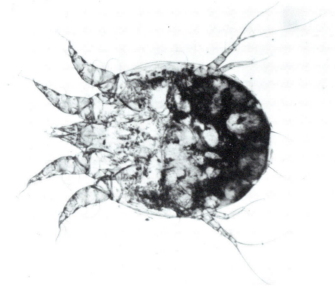

Fig. 2.4.12   A dust mite

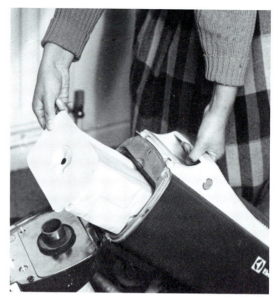

Fig. 2.4.13   Disposable vacuum cleaner bag

Food preservation   In the home it is a problem in hot weather to stop food from spoiling. Butter easily becomes rancid. Milk quickly sours. Meat goes bad and fruit ripens too quickly. Until the invention of the refrigerator much of the food we take for granted had to be eaten very soon after it was collected or bought.

Today, in Britain, about a third of the meat we eat is imported from New Zealand, Australia and Brazil. A journey by ship may take eight weeks, so obviously special equipment is required to keep meat in good condition.

Why does food spoil?   There are several reasons for food spoiling. These include: (1) drying due to loss of moisture, e.g. fruit and vegetables; (2) the food may continue to ripen; (3) it may become rancid by oxidation in contact with air; (4) it may become dirty or spoilt by animals such as rats and insects; (5) it may be affected by micro-organisms such as bacteria and fungi.

Some bacteria produce toxins. These cause food poisoning (see p. 36) in people who eat food containing them.

The most important methods of preserving food and therefore avoiding these problems are concerned with destroying or inactivating the enzymes and micro-organisms responsible for spoilage. They are:

1 Canning
2 Quick freezing and drying (dehydration)
3 Pasteurization
4 Vacuum packing
5 Pickling
6 Curing and irradiation.

Canning   The general principles are (1) to seal the food in a container so that further micro-organisms cannot get in; (2) to heat the container sufficiently to destroy any micro-organisms which are present. See Fig. 2.4.14.

Apart from its efficient destruction of bacteria, one advantage is that no special storage conditions are needed for cans. The disadvantages are weight, involving high transport costs, and that it is only used for food eaten when cooked. Canning does little damage to the nutritional value of foods except destroying certain vitamins.

Fig. 2.4.14 A selection of canned foods

Quick freezing and drying Apart from canning or drying, the most widely used methods of food preservation involve refrigeration. By cooling food below freezing point to −10 degrees C all of the bacteria stop reproducing. Thus decay is very much delayed. When food is frozen slowly, large ice crystals are formed in the cells of the food. The crystals break the cell walls as they grow and when the food is thawed the water drains away carrying salts and other solutes with it. So the food loses its flavour and food value. But the 'quick-freeze' method causes smalled ice crystals to form more quickly. All the moisture in the food freezes before the crystals have had time to reach a large enough size to break the cell walls. When the food is thawed no moisture is lost from it. It thus retains its flavour and its food value and is as fresh after several months' storage as it was at the moment it was frozen.

Dehydration

The word '**dehydration**' means the removal of water. From the earliest times Man has used this method of preserving food. Without water, the micro-organisms which cause food spoilage cannot live. One common practice was to split fish and leave it to dry in the wind. Samples still exist of dried foods preserved 4000 years ago. Industrial methods of dehydration by applying heat to food began in this country in the late eighteenth century.

For many years the products were tough and not very appetizing. Modern methods of dehydrating foods have much shorter drying times. Ways of ensuring that the foods rehydrate successfully (i.e. absorb water again during cooking) have also been improved. Today the shopper can buy a wide variety of dehydrated products which compete in quality with fresh produce, including vegetables, soups, dried milk, cereals and complete meals (see Fig. 2.4.15). The main advantages of dehydration are that the products are light, easy to transport, and if correctly packaged, do not require special storage conditions.

Fig. 2.4.15 A selection of dehydrated foods

Pasteurization **Pasteurization** is a process of applying heat to food and drinks to kill disease-causing organisms. It reduces the bacteria and fungi that cause decay, thus allowing food to be kept for a longer period of time.

The process is named after the French scientist Louis Pasteur. It was developed after his work on the souring of wine and beer had shown that heating them to a temperature of about 54.4 degrees C for a short time delayed souring.

Pasteurization is less severe than **sterilization**. It destroys only a small percentage of the food value and hardly affects the flavour of food. Much higher temperatures are needed in sterilization than in pasteurization. Canned meat and vegetables are sterilized at 115.5 degrees – 121.1 degrees C. The temperature at which milk is pasteurized is 71.7 degrees C. Eighty per cent of a food's vitamin C content is lost merely by boiling it for twenty minutes. At 120 degrees C vitamin B_1 is destroyed. Milk when it is pasteurized loses only ten per cent of its vitamin B_1 and twenty per cent of its vitamin C.

The processing of milk

The best known use of pasteurization is in the preparation of milk. It consists of heating milk to 71.7 degrees C for a fixed length of time, followed by rapid cooling. The heat treatment destroys all the disease-causing bacteria which may be in the milk. At the same time it reduces the number of bacteria which cause milk to go sour. Therefore it will stay fresh for longer.

Milk was first pasteurized towards the end of the nineteenth century. It was then solely used to prevent the milk from souring so quickly. Later it was realized that harmful organisms could be killed if the process were modified. It was found that in order to kill certain bacteria (especially the tuberculosis bacterium) it was necessary to hold the temperature of the milk at approximately 62.8 degrees C for nearly half an hour. This is called the Holder process.

The Holder process has proved too costly because it takes too long and a large unit is required. It has been largely replaced by a higher temperature, short-time method (H.T.S.T.). Here the milk is held at a temperature of 71.7 degrees C for only 15 seconds. This kills the disease-causing organisms and most of the bacteria that cause souring. The milk is then cooled rapidly to below 10 degrees C. If the milk were allowed to cool slowly the milk souring bacteria would increase and rapid souring would take place.

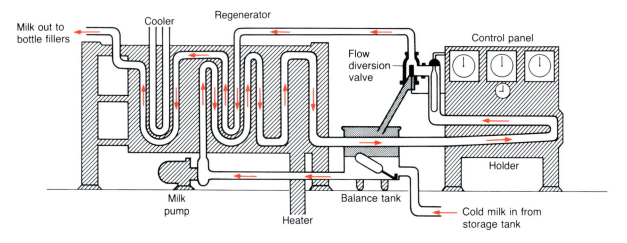

Fig. 2.4.16   Pasteurization

As Fig. 2.4.16 shows the cold incoming milk is first heated by the hot outgoing pasteurized milk. The pasteurized milk loses heat to the cold milk and thus is itself partly cooled. The warmed incoming milk is then heated by hot water in the heater to 71.7 degrees C. Its temperature is carefully kept up until it enters the holding tube. The milk takes only 15 seconds to pass through the holding tube and then it goes to the regenerator where it heats more incoming cold milk. It is piped to the cooler in which it is cooled by brine or chilled water to 4.4 – 7.2 degrees C.

A safety device, the flow diversion valve, automatically opens if the temperature of the milk at the end of the holding period is below 71.7 degrees C and the milk is returned to the heating section. When the required temperature is regained the valve closes and the pasteurized milk passes to the regenerator.

Freshly pasteurized milk is immediately filled into bottles. These have been thoroughly washed and sterilized to prevent reinfection with bacteria. The bottles are capped automatically with aluminium caps.

Vacuum packing

When drying at low pressure, lower temperatures can be used, with less risk of heat damage to the product. In theory then, drying under a vacuum should be ideal. The difficulty is to apply enough heat to the food under the conditions of a vacuum to get a rapid rate of drying. (Some particles are needed for the transfer of heat energy.) Research into vacuum drying has led to the development of accelerated freeze-drying (AFD).

During this method water is changed from a solid to a gas without going through a liquid stage. It is carried out quickly and then the product is dried. The product is then vacuum packed to prevent oxidation.

Pickling

This is the process of soaking food in an acid solution, normally vinegar, where the pH is too low (i.e. the acidity is too high) to allow the growth of micro-organisms or the activity of enzymes. Many vegetables are preserved in this way (see Fig. 2.4.17).

Curing

Food can be 'cured' by the addition of common salt. The more salt is added the more the water in the food is drawn out. Eventually a point is reached at which no more water can be drawn away to dissolve the salt. At this point micro-organisms are no longer able to multiply. Pig products such as bacon and ham are treated by curing. It is also the part of the 'smoking' process applied to fish. Here the smoke acts as a dehydrating agent and the chemicals in it will kill bacteria.

Irradiation

One recent advance in methods of preserving involves neither heating, freezing or addition of chemicals. This is **irradiation**, where food is exposed to sub-atomic particles such as gamma or beta rays or X-rays. The particles or rays destroy micro-organisms but the food develops unpleasant flavours. In addition, the cost of equipment is very high. The most satisfactory results have been produced when radiation is reduced to levels below those which spoil food flavour. The product is not sterile but many of its surface micro-organisms are killed.

Fig. 2.4.17 A selection of pickled foods

Questions: Domains I and II

1 Study the pictures showing various hazardous situations in a home. List the visible hazards in the areas A–K. (II)

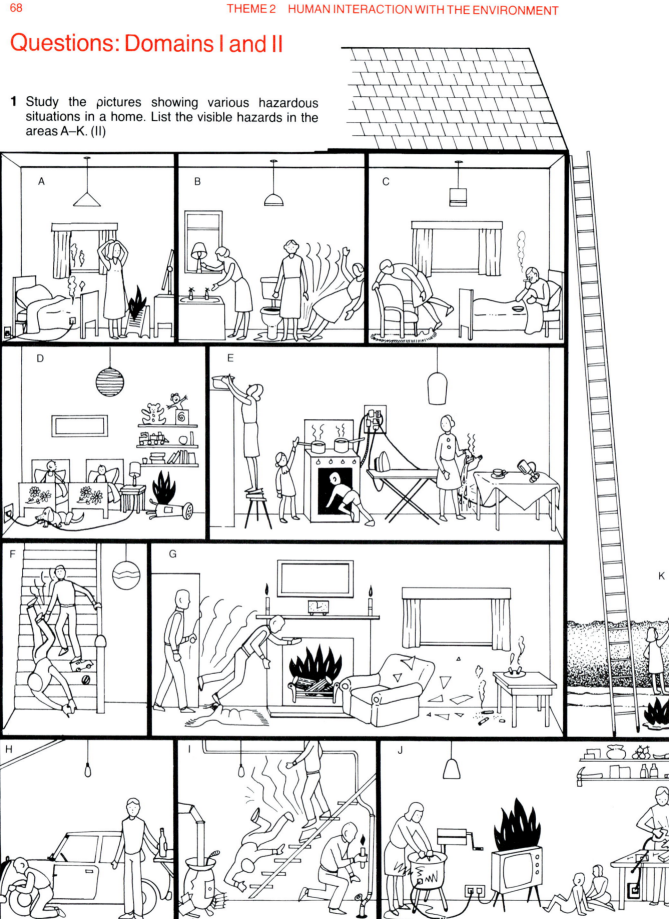

2 Six bottles with screw caps were sterilized, then raw milk was placed into them. They were labelled **A–F**. Bottles **A**, **B** and **C** were then treated by being heated in water at 63°C for 30 minutes, and then cooled rapidly in cold water. Bottles **D**, **E**, and **F** were not treated in this way. **A** and **D** were placed in a refrigerator, **B** and **E** in a warm sunny room, **C** and **F** in a cool dark place. All the bottles were left for seven days.

a Suggest, with reasons, what would happen to the milk in each bottle at the end of seven days. (I)
b Why were the bottles sterilized before being used? (I)
c Name the process described for treating milk in this experiment. (I)
d For each of the following forms of food preservation, explain how it prevents bacteria from affecting the food.
(i) Canning meat. (ii) Drying peas. (iii) Pickling onions. (I)

A – treated refrigerated

B – treated warm sunny room

C – treated cool dark place

D – untreated refrigerated

E – untreated warm sunny room

F – untreated cool dark place

Experimental skills and investigations: Domain III

How useful is disinfectant?

Procedure
1 Take four test tubes and add equal volumes of three brands of disinfectant of equal dilution to three of them. Label them so that they can be identified later. To the fourth tube add an equal volume of distilled water.
2 Cover the base of a Petri dish with 2cm² of nutrient agar. Before the agar sets, add two drops of one type of disinfectant with a pipette. Repeat this for three agar plates using a different disinfectant in each case.
3 To the fourth Petri dish with the nutrient agar, add 2cm³ distilled water.
4 When the agar has set, take a nichrome loop and streak the agar plate with a suspension of soil bacteria in distilled water. Make zig-zags at right angles to each other as in the diagram.
5 Label the plates and incubate them at 25°C for two days.
6 Compare and record your observations on the four plates.

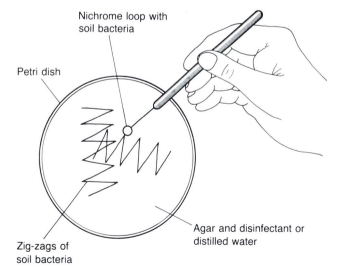

Nichrome loop with soil bacteria

Petri dish

Agar and disinfectant or distilled water

Zig-zags of soil bacteria

Questions
a Which brand of disinfectant appears to be best?
b What was the purpose of the Petri dish with the distilled water?

2.5 We'll take good care of you

Public health It is only in the last 100 years that effective measures have been taken by the community as a whole to combat disease. Earlier attempts had failed because of ignorance of the cause of epidemics and a refusal to learn from the experience of early civilizations such as those of the the Greeks and the Romans.

Public health is the science of preventing disease, prolonging life and promoting physical and mental health. This is the aim of most communities but it has succeeded only in the Western world. The early public health projects of the Greek and Roman civilizations were not copied by the modern world for many centuries. In the eighteenth century, towns in England grew at an alarming rate and their growth continued for a long time. Populations of some towns grew by 78 per cent in ten years. New housing, commonly built back to back and without proper foundations, failed to keep pace. Overcrowding was widespread and serious. As many as forty people lived in houses with five rooms in parts of London. Few houses had piped water; one stand-pipe or well might have to supply a whole street. Sanitation was appalling (see Fig. 2.5.1).

Few people showed concern until the outbreak of a previously unknown disease in Europe – **Asiatic cholera**. The disease reached England in 1831 and was responsible for many thousands of deaths. The government formed a Central Board of Health to advise on treating the epidemic. Two years later, the cholera vanished and the Central Board also disappeared.

Fig. 2.5.1 In the early nineteenth century housing conditions and sanitation were appalling

However, as a result of the outbreak, the Poor Law Act of 1834 was passed. Medical Officers were appointed to give free medical care to the aged, sick and poor and to vaccinate against smallpox. A national survey under the guidance of Edwin Chadwick led to the widespread introduction of improved water-borne sewage disposal systems. Chadwick's work led to the Public Health Act of 1848, which created the General Board of Health. It was in the same year that a second great cholera epidemic swept Britain. This time, evidence was found showing the connection between the disease and drinking water which had been contaminated by faeces. The knowledge quickly led to improvements in environmental hygiene. Cholera was at last controlled and soon disappeared from Britain.

The Public Health Acts of 1872 and 1875 led to the division of the country into 'sanitary areas', each with its own Medical Officer of Health. Poor housing and drainage were to be identified and control was exercised over animal hygiene, refuse disposal, food hygiene and certain trades. Notification of infectious diseases was introduced and isolation hospitals were built. In 1892, Health Visitor training courses began in Britain. The Notification of Births Act of 1907 made births notifiable within 36 hours of their occurrence. Mothers could then be contacted soon after childbirth by Health Visitors.

There was a lot of public alarm at the state of child health. Children lived in appalling conditions in the city slums. Milk depots were opened throughout the country and after the First World War the Maternity and Child Welfare Act of 1918 was passed. It coordinated help to expectant and nursing mothers and their children under school age. The Act enabled local health authorities to provide maternity hospitals and convalescent homes, homes for unmarried mothers, day nurseries, home helps and a district nursing service.

From 1907 onwards, the health of children at school in Britain was supervised by the school health service. The service was staffed by doctors employed by the health authority. Treatment for minor problems was available from 1918. The National Insurance Act of 1911 provided free medical care for lower paid workers. Clinics were opened for the treatment of tuberculosis and venereal diseases.

The quality of life In the nineteenth century infectious diseases were the major killers. The public health movement brought pure water, disposed of sewage and enforced care in the handling of food. As a result, cholera vanished and typhoid and dysentery were controlled. Improved nutrition and housing were followed by the disappearance of typhus and a reduction of tuberculosis. Since then vaccination has almost eliminated diphtheria and poliomyelitis. Vaccines are now available that will protect against measles and rubella. Deaths from infections during and after childbirth are now rare. With the fight against the infectious diseases and the vast improvement in living conditions, the quality of life has altered as well as the length. People now survive hazards of infancy, childhood and childbearing. Expectation of life is 70 years for males and 75 years for females in most developed countries.

In Britain, the Department of Health and Social Security is the central policy-making body in health matters. The minister in charge is an elected politician advised by a staff of civil servants and by a chief medical officer. They administer hospitals and doctors in general practice.

Public health is a responsibility shared with local authorities, e.g. Environmental Health Officers. Each District Health Authority has a local District Community Physician (DCP) who specializes in community health problems. The DCP may be helped by other doctors when an outbreak of infectious disease occurs. Doctors in a local district health authority are also school medical officers, examining children at school. They may be involved in other duties such as the giving of family planning advice. Frequently, medical staff examine (screen) special 'risk groups' such as the very young and the elderly. They often examine large samples of a community who are apparently well for early signs of disease.

In the 1950s Britain began mass X-ray units. They were for chest examination to detect tuberculosis. Mobile units were taken to factories, offices and isolated villages. These examinations showed many conditions besides tuberculosis. Early stages in lung cancer, heart defects and other danger signs were revealed.

Nurses also play an important role in the public health service. There are three types of community nurse: health visitors, midwives and district nurses (see Fig. 2.5.2.). The health visitor's duties include health education, social work and advice. The elderly, mentally ill and the problem family may all be in the health visitor's care.

The district nurse is concrned with nursing the ill within the home. The midwife looks after mothers-to-be during pregnancv and after delivery and may supervise births taking place at home.

Each local authority must appoint a Chief Environmental Health Officer who works with the DCP. Duties include inspection of food and places where food is served such as restaurants; advising on hygiene and often the removal of samples that appear to be unfit for consumption for laboratory testing. Housing is inspected for standards of hygiene to be maintained. The supervision of sewage disposal, drinking water and slaughterhouses is also part of the duties. Other responsibilities include control of air pollution, noise levels, and hygienic accommodation in shops and factories.

Many health authorities now employ Health Education Officers. Dental officers provide dental care for expectant mothers and young children. Mental health officers, social workers and home helps now form part of the Social Services departments.

Fig. 2.5.2 Community nurses: (a) health visitor; (b) midwife and (c) district nurse

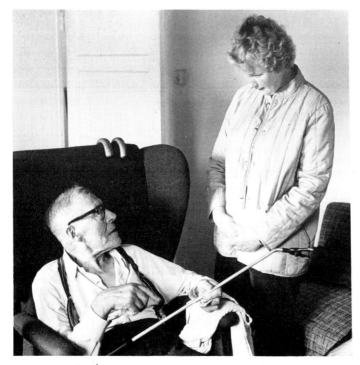

(a) △

◁
(b)

(c)
◁

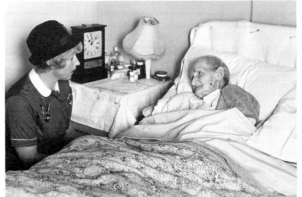

New problems for old In the nineteenth century, death in childhood was so common that it caused little comment. Expectation of life was as low as 16 years for children of labourers living in towns. The public health measures of a hundred years ago aimed at cleaning up the environment and started to reduce infant mortality. However, even in 1910 about 120 babies out of 1000 would die before their first birthday. A common cause of death was summer diarrrhoea, the result of food contamination. Then health visitors began to bring advice. Mothers became better educated. Infant welfare clinics and milk depots offered help and clean milk. Infant mortality fell quickly and has continued to fall. In the poorest countries, public health measures may be confined to campaigns to eliminate malaria, smallpox, and poliomyelitis. The World Health Organization (WHO) plays an important part in this sort of activity.

With the decrease in infant mortality has come the smaller family and a falling birthrate in Britain. Better health care means surviving to old age. But this can bring new hazards. People over retiring age make up an increasing proportion of the total population. With ageing come the the 'wear-and-tear' diseases and mental stresses that are a part of modern life. Hence the new problems facing public health are those of high-pressure living and of old age (see Fig. 2.5.3).

(a) △

(b) △

Fig. 2.5.3 The new public health problems: (a) high-pressure living and (b) old age

Coronary artery disease and cancers are now the leading causes of death. Over-eating, smoking and lack of exercise are all suspected as causes and health education is attempting to alter attitudes to these behavioural problems. Air pollution and smoking are largely to blame for emphysema and bronchitis. Keeping a check on pollution is an increasingly important function of public health.

Prevention or early detection of disease is better than late attempts at a cure. Cervical cytology clinics seek out women at risk from cancer of the womb. Genetics clinics may offer advice to couples with abnormal babies, who want to know the chances of having other affected children. Genetics clinics may offer advice to couples before they start a family because couples often want to know the chances of having children with inherited disorders on the basis of past family history.

While the developed countries have such systems, it must be remembered that the majority of the world's popluation live in conditions of poverty, overcrowding and semi-starvation. The greatest public health problems to be tackled on a world scale are still caused by parasites, malnutrition and over population.

Water supply

Water supply is an essential feature of everyday life. It is often taken for granted in cilivized communities. Water authorities are required by law to provide pure water. It must be free from visible suspended matter, taste and smell, and from anything which could be harmful to health. In a typical town more than fifty gallons of water are provided each day per person.

Water is obtained from rivers. Some areas have wells which provide much of the water required. Wells in the London area provide 15 per cent of the Metropolitan Water Board's water. The rest comes from the Thames and its tributaries. River water is usually muddy and contains millions of harmful bacteria. All have to be removed before the water is considered pure enough to pass through your kitchen taps. Many towns have no suitable river from which they can obtain water. In such cases water will have to be piped to the town's water works. Sometimes a whole valley may be dammed to produce a semi-natural lake which will act as a reservoir. This has been done in Cumbria to provide water for Manchester, many miles away.

When water is pumped from the river it goes into a large open **reservoir** (see Fig. 2.5.4, stages 1–3). This serves to help out the water supply during dry periods. It also plays a valuable part in the **purification process** as a large amount of suspended material settles here. Unfortunately, the conditions in a still body of water are ideal for the growth of minute plants called algae. The plants are removed by **filtration** but the filters may require more frequent cleaning if algal growth is heavy. Chemical treatment of the reservoir water may be used to reduce algal growth. Care must be taken at the reservoir to prevent any disturbance of the sediment and to avoid floating debris getting into the filters. At the **filtering station** (stages 4 and 5) the water is allowed to trickle down through layers of sand and gravel. Treatment of the water, with chemicals which cause the particles to clump together, increases the efficiency of filtration but is not always practical.

The **primary filters** (stage 4) are of coarse sand and are easily cleaned by blowing currents of air through them. The water runs through this type of filter at a speed of about seven metres per hour. The partially filtered water obtained is fed onto the surface of the **secondary filters** (stage 5). These consist of gravels covered with about half a metre of fine sand. Under the gravel are porous tiles forming drains through which the water passes to the next stage. These secondary filters are much slower in their action. The water moves through at under thirty centimetres per hour. This is because the fine sand has much smaller spaces through which

Fig. 2.5.4   Water treatment

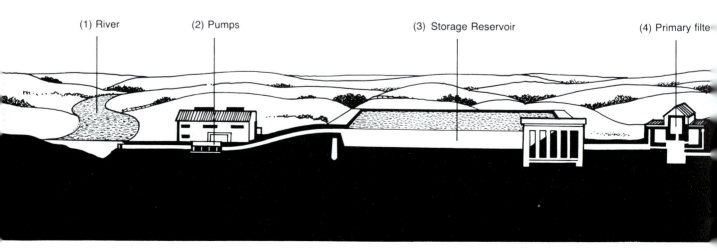

(1) River   (2) Pumps   (3) Storage Reservoir   (4) Primary filte

water can flow. This system is, however, preferable to one where only a fine filter was used which was always being choked with large particles.

As the process goes on, a film of fine silt, debris and micro-organisms builds up on the surface of the sand. The bacteria break down complex organic matter into harmless inorganic compounds. The bacteria multiply to form a jelly-like layer round the sand grains in the surface layers. Suspended material is caught on this jelly and provides the bacteria with food. However, this deposit makes a larger head of water necessary to maintain the required flow through the filter. Eventually it becomes necessary to shut off the water supply and clean the filter by removing the top inch or so of sand and washing it in clean running water. The washed sand can then be replaced and the filter returned to normal use. There are several of these filters at the water works so that each can be cleaned in turn without too much hindrance to the purification process.

Some minor operations may be necessary, according to local water conditions. For example, iron, which would give the water a sour taste and would stain fabrics during washing, must be removed. Other minerals are found locally and are removed by chemical processes which will vary according to the mineral. After removal of these, the water passes to the **chlorination plant** (stage 6) where it is treated with chlorine gas. This kills bacteria in the water. Continuous testing of water passing out of the plant ensures that the chlorine level is not high enough to make the water distasteful.

Although these purifying processes are very efficient, the water is still tested daily for bacteria. A harmless bacterium, *Escherichia coli*, found in everyone's intestine, is very hardy and difficult to kill. It is an indicator of purity. If only very few of these are present in a test sample it can be concluded that the water is safe to drink.

Large pumps pass the pure water from the pumping station (stage 7) to a service reservoir (stage 8). This is a covered chamber where the pure water is stored. It passes from here through pipes that supply houses (stage 9). The best site for a service reservoir is high up so that the water will build up pressure to run freely from the main pipes. A large town will have a number of these reservoirs, each supplying a certain area. Small hill top areas often have storage towers. Branch pipes leading to houses have large taps or valves at regular intervals. These are needed for tracing leaks and shutting off the water from a small area so leaks can be repaired. Opening kitchen taps causes the water to flow out of the pipe and down the drain – eventually reaching the river to begin the cycle again.

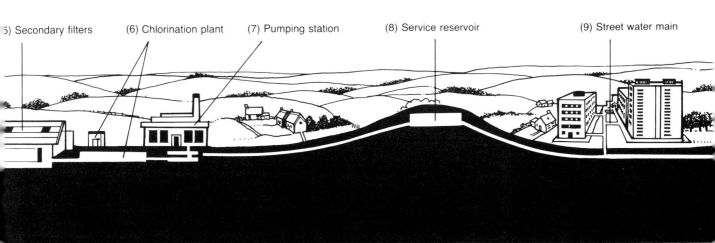

(5) Secondary filters   (6) Chlorination plant   (7) Pumping station   (8) Service reservoir   (9) Street water main

Sewage

Sewage is Man-made waste. Great amounts of it are produced by all of us. It includes the solid and liquid wastes from out bodies – faeces and urine. Sewage also contains the waste water that people throw out after washing and some industrial waste from factories.

Sewage must be carefully disposed of. It smells foul and can be dangerous. Sewage contains harmful organisms, such as certain bacteria. These may be carried onto food by rats, insects or unclean hands. They may pass into drinking water from sewage dumped in rivers. Also, they may drain into rivers from sewage dumped in soil. If they are eaten or drunk, they could be a serious health hazard.

The word, sewage, comes from the Latin verb meaning to drain away water. This is how the earliest methods for disposing of sewage worked. Underground pipes were used to carry sewage from houses to deep pits. The pipes were 'flushed' by pouring water through them. Such systems were built as long ago as 1500 BC.

As cities grew bigger, so did the amounts of sewage. By the nineteenth century sewage pits were enormous. Harmful bacteria grew on the sewage materials and spread into city water supplies. People often died of diseases, such as cholera, after drinking contaminated water. Modern methods dispose of sewage safely.

Today, most cities have strict laws of sanitation. This word comes from the Latin, meaning health. Untreated sewage can no longer be legally dumped into open pits or poured into lakes and rivers.

In most British houses, pipes carry sewage from the sink, bath and toilet. The pipes run into a system of underground drains called **sewers**. A sewer is usually laid under each street for the houses above it. These local sewers from several streets lead into a larger, principal sewer and so on (see Fig. 2.5.5, stage 1). Where possible sewage flows downhill.

About 400 litres of sewage are produced per person per day in a large city. Only about a half a litre of this is solid matter. These solids cause most of the health risk and have to be made harmless. There are several stages of treatment.

First, the sewage is passed through **wire nets** or **screens** (stage 2). They collect big objects such as pieces of wood. If not removed, these would damage the pumps. They may be burned or ground into small pieces. Then the sewage flows through wide pipes called **grit channels** (stage 3). Here the grit and gravel is collected. It is relatively heavy and settles to the bottom. It may be buried or washed and used in industry.

Most of the remaining solids in the sewage are organic matter. It is pumped into large tanks where it stays for several hours (stage 4). Much of the solids settle to the bottom. Such a process is called **sedimentation**. A solid settlement called **sludge** is formed.

Up to this point the sewage has had only **primary treatment**. Some cities do little more with their sewage. They are built on rivers or by the sea. They pump sewage, minus the sludge, into this water. It is soon diluted, or thinned out, by the water. Chlorine is added to it which kills harmful bacteria. The danger from the sewage thus becomes very small.

Secondary treatment may be needed in other cities. It is used where small rivers or streams would be endangered by the sewage. Its main device is the **trickling tank** (see Fig. 2.5.5, stage 5). This contains a bed of stones about 180cm deep. The stones are up to 10cm wide. On their surface is a slime, made of micro-organisms. These are mainly 'helpful' bacteria which live on dead matter. Sewage drops onto the bed of stones from a pipe. It trickles down through the stones. The bacteria on the stones digest the organic matter in the sewage. They break it down to harmless chemicals.

Next, the sewage passes to another tank (stage 6). Remaining solids in it can settle out. By now up to 95 per cent of the solids in the sewage have been removed. Then chlorine is added to kill any remaining organisms (stage 7). The sewage can then safely be pumped into streams.

Finally, the solid sludge collected during the treatment must be disposed of. It is put in large **digestion tanks** (stage 8). These are heated to speed up the action of bacteria in them. After about 30 days, the bacteria have broken up the sludge into harmless substances. The purified sludge is then dried on beds of sand and gravel (stage 9). The dried sludge may be buried or sold as fertilizer.

A house in the country may not have any connection to sewers. It often has its own **septic tank** to dispose of its sewage. The sewage flows to an underground tank where solids settle to the bottom. The sludge is broken down by bacteria. The liquid can overflow into another tank with bacteria. Thus the liquid is purified and drains into the soil. The solid sludge is cleaned out of the septic tank at intervals.

Fig. 2.5.5 Sewage treatment

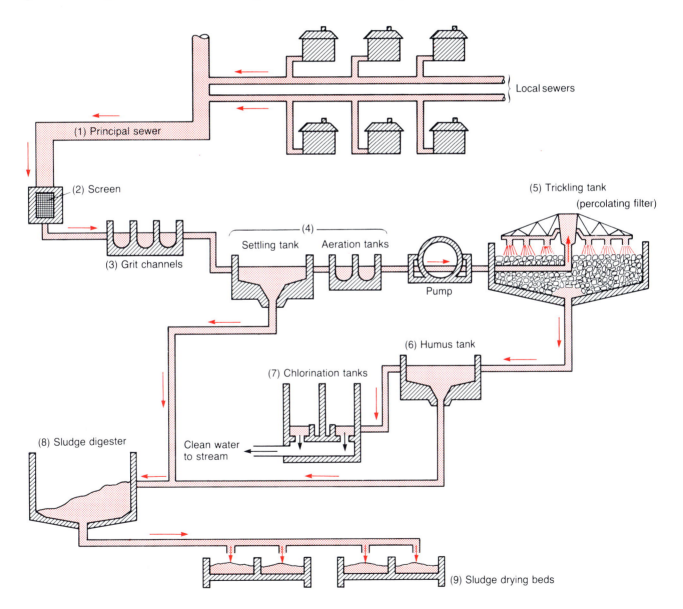

Questions: Domains I and II

1 Read the following passage and answer the questions that follow.

It is considered that the provision of an adequate supply of water fit for drinking and cleaning purposes is the most important requirement of any community. In Britain an average supply of 30–40 gallons (140–$180\,dm^3$) of water is required for each person, each day. Only a small percentage of this is required for drinking, the remainder being used for cooking, personal washing, cleaning and flushing toilets.

The provision of such large quantities of clean water presents various problems, such as where these large amounts may be found, how the water can be purified and how it can be stored safely and delivered to the community. Some of the pollutants which may have to be removed, depending on the source of the water treated, are sewage, micro-organisms, industrial wastes, detergents and suspended clay.

Water for supply may be obtained from natural or artificial lakes, such as those in Wales which supply Liverpool and Birmingham; rivers, e.g. London obtains much of its supply from the Thames; and deep wells, which are also used to supply London. These wells are 15 metres deep or more. Some small communities still use water from shallow wells which may be less than eight metres deep. The water in such wells may easily be contaminated and they should be very carefully investigated before they are used. The water in them may be polluted from soil and vegetation entering at the top, surface water flowing into them during heavy rain, and dirty water entering at their sides because it has drained through a thin layer of subsoil which is insufficient to filter out the impurities. The water obtained from shallow wells may be made safe by careful construction of the well, such as providing it with a close-fitting air-tight top and by lining the top six metres of the sides with a waterproof layer.

Water from a deep well is often very pure because the rain water supplying it has had to filter through a great depth of subsoil and rock. The only treatment which it might require is the addition of chlorine. This ensures that harmful micro-organisms, which may enter during storage in reservoirs and transport through pipes, are killed. However, not all deep-water wells are so pure. Cracks in the rocks around them may allow surface water to enter without adequate filtration.

Water from lakes and rivers is more heavily polluted, and so more difficult and expensive to purify to the standard required for public supply. Water from lakes may contain algae, which have grown because the water may contain nutrients suitable for their growth, and the still water helps their formation. Some of these algae may be poisonous, others may produce an unpleasant taste and smell. They are difficult to remove during the treatment of water.

Such waters must be treated by sedimentation. The water is allowed to stand for a time. This allows much of the matter suspended in the water to sink and it can account for 90 per cent of the purification required. Chlorine is added to the water in an amount sufficient to kill micro-organisms, particularly those which might cause disease. Flocculants such as iron sulphate or aluminium sulphate are added. These substances cause the very small particles floating in the water to stick together so that they settle out under gravity. Filtration through sand or similar material removes the last suspended matter. The water is then suitable for drinking.

Water delivered for public use still contains a small amount of chlorine (about 0.5 parts per million). Although at this concentration the chlorine is tasteless and odourless it protects the water during storage in covered reservoirs and passage through pipes to the home. With increasing demands for water it is possible that another major source of water for treatment will be used. At present much of the water contained in the effluent produced by sewage treatment is discharged into the nearest river or into the sea. It is possible that this water will be retained in the water reclamation works, which receives the sewage, and purified so that it is once again fit and safe for drinking. It could then be returned to the water-supply system.

Questions
a Name the five sources of drinking water which are used in Britain. (II)
b Name the city which obtains its water from two sources. (II)
c What is another possible source of water for household use which the passage suggests may be used in the future? (II)
d Explain why water from a shallow well may be safe to use providing the well is covered with a good-fitting lid. (II)
e Why will lining a well with a waterproof layer improve the quality of its water? (II)
f What word is used in the passage to describe a substance which causes small particles to stick together? (II)
g Why is water obtained from a deep well often very pure? (II)
h Suggest three reasons why surface water is more expensive to purify than deep well water. (II)
i Name three pollutants which may be present in untreated water and for each pollutant suggest one reason why it must be removed in order to make the water drinkable. (II)
j What is the purpose of maintaining a small amount of chlorine in the water sent for use in homes? (II)

2 The plan opposite shows the main stages in the treatment of sewage.

a In which part of this process are the following used as important parts of the treatment?

　i Aerobic organisms.

　ii Anaerobic organisms. (I)

b Give two important differences between the solid material entering at A and leaving at B. (I)

c Name the gas leaving at C. (I)

d To what use can this gas be put? (I)

e What use may be made of the solid leaving at B? (I)

f Give two reasons why the supplying of clean water to homes is important. (I)

g Frequently water from lakes or rivers is purified to provide a town's water supply. Name two contaminants which must be removed from such water to make it fit to use. (I)

h In the process of water purification what is the purpose of:

(i) adding chlorine; (ii) storing in a covered reservoir? (I)

3 Study the drawing of a grocer's shop with adjoining toilet and refuse provisions.

a List six ways by which the food in the shop might become contaminated. Study the complete drawing and list six unhygienic conditions. (II)

b Suggest six remedies for improving the hygiene standards of these premises. (I)

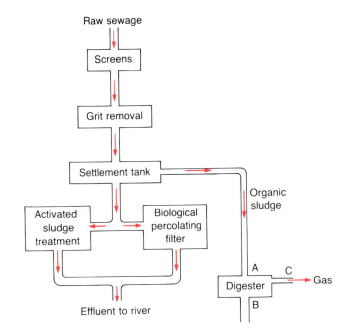

c How should the following items of household refuse be kept ready for collection?

Cardboard packets, bottles, cans, food scraps. (I)

d How does the local authority dispose of general household rubbish after collection? (I)

e How would you dispose of an old refrigerator and some garden refuse? (I)

2.6 Live now pay later

Pollution

For hundreds of years, the human species has increased its population and has treated the planet Earth as a bottomless sewer. Only recently have we become aware of the massive problem that we have helped to create.

Earth is three-tenths land and seven-tenths water. Only one-tenth of the land is cultivated, so this is the extent of our environment. Throughout history, mankind has become expert at introducing unwanted materials into this environment, thereby polluting it. Even the air we breathe and the water we drink has not escaped misuse.

City life and death

Thousands of years ago, overcrowding began to cause slum conditions with the problems of spread of disease, sewage disposal and rubbish collection. We know this because archaeologists have found miniature versions of present day pollution problems in the remains of cities long buried.

Every day in modern cities, millions of plastic containers, cans and bottles are thrown away. We produce more refuse in one week than ancient civilizations produced in the whole of their history.

As long ago as 1000 BC there were attempts at making sewage systems and water supply pipes. Dry refuse was put in pits outside cities and allowed to rot into compost for manuring fields. In AD 100, Rome had nine aqueducts carrying clean water to the city. Lead pipes carried water to a thousand public baths and houses. A main sewer passed into the river Tiber which flowed into the sea.

Today people are very concerned about the pollution of the Mediterranean by untreated sewage and wastes from factories.

Air pollution has had a similar history (see Fig. 2.6.1). Coal has been mined and burned in Britain since the thirteenth century. By the time of Elizabeth I cities were already being blackened with soot. Fog was a common cause of death in Dickens' London, and in this century it killed thousands of people. In 1952 as many as 4000 died in one London smog (a mixture of smoke and fog). A Clean Air Act, insisting on smokeless zones, enabled the buildings in London to be scrubbed back to their clean state. However, this was too late for the victims of previous years (see Fig. 2.6.2).

Fig. 2.6.1 Smog over Los Angeles

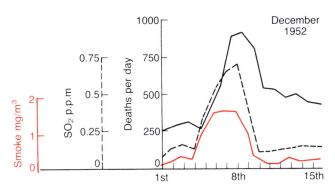

Fig. 2.6.2 Graph of death and pollution levels in the London smog of 1952

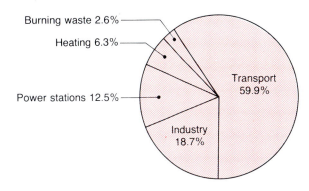

Fig. 2.6.3 Percentage composition of air pollution in an industrialized town

The age of the internal combustion engine added to this form of pollution. Today, 6000 million tonnes of carbon dioxide pass into the atmosphere each year (see Fig. 2.6.3). Much of this comes from car exhausts and is a public health concern. It can even affect the climate.

Mankind's survival has, unfortunately, depended on our wrecking much of our environment. We have cut down forests for timber. We have changed the course of river systems, quarried rocks and made deserts by causing soil erosion. Nature also played a part. Over thousands of years volcanoes added gases to the atmosphere, earthquakes have released oil and natural gas, and rivers have washed out poisonous metals into the sea.

Some of these poisonous metals are used today in industry. Their concentration tends to build up in living things and this causes concern when we eat them. Particular blame is attached to paper making. Mercury was used to destroy fungi which interfere with paper making. It gets into the waste that is discharged from paper mills and is also released in smoke from burning paper. In addition it is used in pesticides for agriculture. When the mercury enters water it is built into the bodies of microscopic plants. These are eaten by animals such as fish which continue to build up the concentration through the food chain (see p. 26). Eventually, animals at the end of the food chain will be poisoned.

If more and more people want more and more cars, then oil-tankers will become larger and larger. The risks of collision, spillage and oil pollution are increased. If hundreds of millions of people want more lather in their laundry, then drains, rivers and sewage works will have to cope with detergents which are difficult to break down. If they want plastics, then they are likely to have fumes and end products which will last for many years. If firms want quick profits and people are greedy for products they never previously knew they needed, then industry will throw away the waste materials and create more pollution.

More people to destroy In our attempt to keep more of our own species alive, we have increased pollution. We have discovered the insecticide, DDT, and used it to destroy malaria-carrying mosquitoes thus saving millions of lives. However, our world still suffers from many insect-borne diseases despite our efforts. Malaria, plague, typhus, sleeping sickness, and many others led to an enormous wastage of life. DDT, an organic chlorine compound, is now considered as a pollutant and severe restrictions have been put on its use by many countries. Like the poisonous metals mentioned previously, it builds up in food chains, causing problems in carnivores.

By AD 2000, seven thousand million people may have to live together on our planet. Already mankind has perfected the art of making war. When atmospheric bomb testing was considered necessary, countries with nuclear bombs had their test sites as far as possible from their own lands. They claimed that the radioactive fall-out would be safely localized. It was confidently thought that the dangerous products would be diluted in the atmosphere. Disaster struck as these products swept around the world in air currents. Every person who grew up when bombs were being tested in the atmosphere has in his or her bones, radioactive strontium. The after-effects of prolonged exposure to radiation were only witnessed many years after.

Nature's persistant progress pollution lasts!

DDT is found in the flesh of penguins in the Antarctic, where no DDT has ever been used. It could only have got there from northern farmlands, carried by rivers into the ocean currents, or by airborne dust. Mercury occurs in tuna fish – a reminder of all the mercury which has reached deep sea fish.

In Britain, the authorities claimed that they had reduced the amount of sulphur dioxide to insignificant proportions. The Norwegians disagree and claim that Britain's sulphur dioxide in the form of acid rain is attacking their forests, hundreds of miles away.

In the long term future we can foresee even more terrible effects. The atmosphere is the envelope which protects living things from harmful radiation from the Sun and outer space. It controls climate. Millions of years ago the Sun helped the growth of forests which became our coal. The plant growth in the sea became our oil. Those fossil fuels led to pollution from chimneys and exhaust pipes. This produces smog which affects people's lungs. The carbon cycle in nature is a self-adjusting mechanism. Carbon dioxide is essential for plants. It is a source of life but there is a balance which is kept by excess carbon being absorbed by plants in the seas. The excess is now too much for this absorption. It can seriously disturb the heat balance of the Earth because of the 'glasshouse effect'. A glasshouse lets the Sun's rays in but retains the heat. Carbon dioxide does the same. It keeps the heat at the surface of the Earth and when in excess can change the climate. It has led to ice-caps and glaciers melting and an increase in the average temperature of the sea in some parts of the world.

A disastrous blunder

In Pakistan the population is increasing at a rate of two per minute. In the same time almost an acre of land can be lost through water-logging and increasing the amount of salt in the soil. This is the largest irrigated region in the world. Twenty-three million acres are watered by man-made canals. The British began to farm this region in the nineteenth century. The land was fertile but had a low rainfall. Irrigation canals were constructed but they were not lined. The water simply soaked into the land instead of spreading out and draining back into the Indus river. Some 40 per cent of the water was lost underground. This resulted in a raising of the water table, drowning the roots of crops. In other areas the water crept upwards reaching salts which collected at the surface, killing the crops. Today a white crust of salt glistens like snow on vast areas of useless land. It would take many years and thousands of millions of pounds to repair the damage.

Unthinking we treat rivers like sewers and lakes like cesspools (see Figs 2.6.4 and 2.6.5). These natural systems have struggled hard. Microbes which cope with reasonable amounts of organic matter were destroyed by detergents, sewage and waste from factories (see Fig. 2.6.6). We have had the 'Freedom from Hunger Campaign'. Presently, we may also need a 'Freedom from Thirst Campaign'. If something is not done about protection of the purity of fresh water, we will face a desperate

Fig. 2.6.4 This mid-nineteenth century illustration appeared in the magazine *Punch*, entitled 'The silent highwayman – your money or your life'. It was an attack on the public indifference to the polluted condition of the Thames, a cause of many epidemic diseases including cholera

Fig. 2.6.5 *Punch* cartoon – 'Old Father Thames'

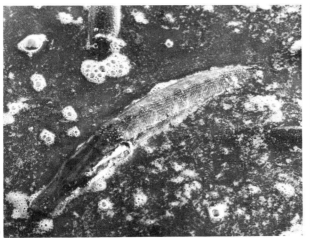

(a)

(b)

(c)

Fig. 2.6.6 Water pollution can take various forms: (a) dead fish; (b) sea-birds covered in oil and (c) effluent from factories

situation. The competition is between the water needs of the land which has to feed us and the population's industrial needs. The least amount of water that each person needs per day, in advanced industrial countries, is about 300 gallons. This is just a drop in the ocean of water needed by steel-making, paper-making and the chemical industry. To die of hunger one needs more than 15 days. To die of thirst one needs only three. Yet we are wasting, polluting and destroying water every day. In many countries more water is being used than is falling as rain. Africa can testify to this (see Fig. 2.6.7).

Fig. 2.6.7   In Africa the aim is 'clean water for all' so that people do not have to walk miles to collect water or drink from contaminated streams

Pollution is a crime to humanity. It begins as the greed of the producer wanting to make a quick profit and of the customer who wants many things, regardless of the cost. The customer then becomes the 'public' singing protest songs about pollution. We are not totally ignorant of the effects of destroying the purity of our environment. However, there is still much to learn about how pollutants are transported by air and water. This means monitoring on a world scale and creating international standards about what is permissible. Perhaps it is already too late.

Conservation

A balance in nature may remain undisturbed for thousands of years except when altered by natural disasters such as volcanoes, earthquakes or floods. Unfortunately, Man has it in his power to upset the balance of nature. He also has the knowledge and ability to restore, by slow degrees, what he has destroyed.

Ecology is the study of the inter-relationships of plants and animals in their natural habitats. It provides guiding principles by which we can conserve natural resources and thus restore the successful balance of nature. For thousands of years Man has protected those wild animals he has used for food and sport. It is only in the twentieth century that he has tried to preserve plants and animals for their own sake and for future generations (see Fig. 2.6.8). Today there are many national laws protecting forests and wild flowers and providing refuge for wild animals. International agreements also exist to preserve many animals such as whales and seals from extinction.

Fig. 2.6.8   A 'Save the seal' demonstration

Forests   Forests are extremely valuable natural resources. Trees may provide a variety of products including building materials, pulp for rayon and paper, resins, dyes and methanol. Forests also provide natural places to live for so many forms of wild life. They prevent soil erosion and flooding. They are also very beautiful and this, in itself, is a good enough reason to ensure their survival for future generations (see Fig. 2.6.9).

Well-planned conservation programmes have been set up by forestry commissions in most countries. Their aims include:

Wise forest management by careful pruning, weeding and thinning of trees.
Renewing trees which have been removed or killed.
Protection of tree seedlings.
Prevention of plant disease.
Limiting removal of trees.
Prevention of forest fires.
Education of the public to appreciate the aims of conservation.

Fig. 2.6.9   Forestry conservation

Fig. 2.6.10   A nature reserve – Tsavo Park, Kenya

Fig. 2.6.11   Wild flowers are beautiful and should be preserved

Wildlife

The controlled use of wildlife is very important. It is in our interests to protect it. Sea fish and invertebrates are used extensively as food. Some fish produce essential oils, others fertilizer. Useful insects destroy pests which damage our crops. Birds destroy harmful rodents and insects. We obtain meat and skins from certain wild animals. For example, the beaver is useful for its fur and it also prevents soil erosion by building dams. Many fur farms breed silver foxes, mink, muskrat and beaver for commercial reasons and serve to prevent possible extinction of these species. Some of the laws which have been passed to protect wildlife include closed seasons for hunting and limiting exploitation of animals during their breeding seasons.

The minimum size and the maximum numbers of many species are specified. Nature reserves and National Parks have been set up throughout the world in which the killing of animals and plants is not allowed (see Fig. 2.6.10). For example in Scotland such a reserve protects arctic-alpine plants and the golden eagle. In Suffolk a reserve protects the avocet. Huge areas of land have been set aside in Africa for the protection of many species of game animals. The Australian government has set up reserves to protect the unique animals and plants of Australia.

Besides setting up land nature reserves, governments have gone some way towards setting up marine nature reserves in some parts of the world. Fish hatcheries help to replenish stocks of many species of both marine and fresh water animals. Indeed many public lakes and rivers are stocked from such hatcheries with trout and salmon.

It it only relatively recently that wild flowers have been recognized for their beauty. Their possible extinction has become alarming and organizations such as the Botanical Society of the British Isles have been set up for their protection (see Fig. 2.6.11).

Soil Without soil Man would find it impossible to grow his crops. **Erosion** of soil has been taking place for million of years by the action of wind and water. Top soil is carried from one area and deposited in another. Some parts of the world have benefited by this depostion of soil, for example the Nile Delta and the Mississippi Delta. Other areas show the opposite effects and remain as dust bowls. Where there is thick vegetation, the roots bind the soil together and prevent erosion. When forests are destroyed therefore, erosion is encouraged. Fields in which animals have been allowed to graze until the land is bare of grasses are subject to erosion by wind and rain (see Fig. 2.6.12).

Top soil can become depleted of minerals making it useless for growing crops and more likely to be eroded. Poor management of soil has resulted in wasting vast areas of soil. Conservation has taken the form of scientific farming methods. These include:

Planting of grasses and shrubs which hold soil particles together with their roots.
Planting in lines at right angles to the slopes of hills to prevent erosion by run-off rain water.
Replanting of trees around sources of rivers.
Crop rotation.
Addition of fertilizers.
Addition of lime.

Fig. 2.6.12 Soil erosion – the Dakota Dust Bowl, USA

Questions: Domains I and II

1 The bar chart shows the estimated amounts of pollution in the atmosphere in a country in one year.

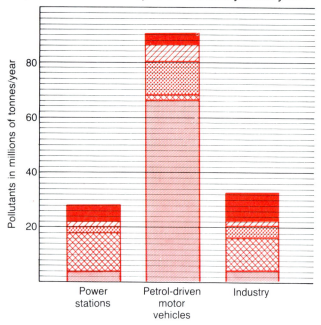

KEY

- Particles
- Nitrogen oxides
- Hydrocarbons
- Sulphur dioxide
- Carbon monoxide

a What is the total mass of pollutants from industry? (II)

b Which source of pollution produces most particles? (II)

c Make a table as follows and complete it to show the amount of nitrogen oxides and hydrocarbons produced by power stations and petrol-driven vehicles. (II)

	Power stations	Vehicles
Nitrogen oxides (tonnes per year)		
Hydrocarbons (tonnes per year)		

d It has been suggested that all petrol-driven motor vehicles should have exhaust purifiers. What evidence in the chart supports this idea? (II)

e Name one of the pollutants on the chart which is dangerous to health and state why it is dangerous. (I)

2 The Wabigoon river in Canada was an attraction for tourists and anglers. A paper mill discharged its waste mercury into the river until 1975. In 1969 an analysis of fish from the river had shown that they had 12 to 24 parts per million of mercury in them and that the livers of fish-eating birds from the same river had 100 p.p.m. of mercury. The plant plankton had 0.5 p.p.m. of mercury in their cells.

a Using the above information, construct a food pyramid (pyramid of numbers). (II)

b Explain why there is more mercury in the fish-eating birds than in the plankton. (I)

c Local Indians had high levels of mercury in their blood and showed symptoms of poor vision, hearing defects and abnormal muscular contractions.
Suggest:
how the mercury reached their blood;
which tissue, besides blood and muscle, is affected by the mercury. (I)

3 Between 1958 and 1968, the reproduction rate of the rare and protected Bermuda petrel (a type of carnivorous sea bird) declined by 3.25 per cent. The lowering of the reproduction rate is linked to the use of DDT. The Bermuda petrel feeds entirely at sea. Concentrations of DDT, from five unhatched eggs, averaged 6.44 parts per million. Chicks often died before, or very soon after hatching.

a Name the class of vertebrates that forms the main diet of the Bermuda petrel. (I)

b State what DDT was used for. (I)

c The concentration of DDT in sea water was much less than 6.44 p.p.m.
i Suggest how DDT reached the water. (I)
ii Explain how DDT reached a concentration of 6.44 p.p.m. in the Bermuda petrels. (I)

d Suggest why the Bermuda petrel needs to come on to land. (I)

4 As a result of heavy traffic on motorways, the central verges have become polluted. Despite this, herbivorous insects such as the larvae of the buff-tip moth and the gold-tail moth thrive on the vegetation growing on the motorways' central verges. Both species feed on leaves and are normally eaten by birds when they occur away from the motorways. Pollutants such as oxides of nitrogen in the atmosphere and sodium found in plants, occur in unusually high concentrations on the motorways' verges.

a Name the major class of food that can be increased in plants by increasing the intake of nitrogen. (I)

b Suggest another pollutant from traffic, besides oxides of nitrogen or sodium, which could affect organisms living near the motorway. (I)

c From the account given, state two reasons why herbivorous insects thrive on plants near the motorway but not on similar plants growing far away from the motorway.

d During winter, de-icing salt is often used on the motorway. Explain how this could affect the ability of the plants to obtain water. (I)

5 The following article appeared in a magazine in 1979, when it was feared that the Large Blue butterfly had become extinct in Britain. Read the article and answer the questions which follow.

The End of the Large Blue Butterfly

After feeding on the flowers and developing seeds of wild thyme for three weeks, the Large Blue larvae drop to the ground where they are discovered by feeding ants. The larvae have glands producing a sweet substance that attracts the ants; after feeding on this substance the ants carry the larvae into their nest, where the larvae eat ant grubs. The larvae hibernate during winter, when half-grown, and resume feeding in spring. They change into adult butterflies in June and feed on nectar.

New Scientist

a Name two organisms which have to be conserved to prevent the extinction of the Large Blue butterfly. (II)
b At what two points in its life-cycle is the Large Blue a primary consumer? (I)
c At what point is it a secondary consumer? (I)
d Name the producer in the life-cycle. (I)

6 The following article appeared in a newspaper. Read it and answer the questions which follow.

Why Frogs Have Hopped It

The main pond in the village for spawn was close to the brook. After the Anglian water authority 'cleaned' it out, the water table fell. For a few years the dwindling number of frogs tried to breed in the brook but failed. The village pond was once used by farmers to water their horses and to wash the wheels of their carts.

Just a few years ago it was the summer home of numerous newts. They too have declined, possibly because of salty water washing in from the roads. The other ponds in the parish are also frogless.

There is some hope that frogs will return to the village. Last summer the water authority dug the old brook pond deeper and it now holds water once more. Spawn for the restored pond came from several sources: from a garden pond in a neighbouring village, from an old pond that was to be filled in and from a flooded field several miles away. Already hundreds of tadpoles have hatched and later, I hope, the small frogs will successfully leave the water to fend for themselves.

a State:
  i a characteristic of the class amphibia which is mentioned in the article;
  ii a change in Man's farming methods which has helped in the decrease of frogs;
  iii a reason why the frogs failed to breed in the brook. (II)
b i Name the pollutant which is probably killing the frogs.
  ii Why is the pollutant used by Man? (I)

c Explain concisely how this pollutant kills amphibians. (I)
d State two ways, mentioned in the article, that Man is trying to conserve frogs. (II)

7 In 1986 an accident at the Soviet Union's largest nuclear power station in Chernobyl released a cloud of radioactive material high into the atmosphere. The winds then blew this cloud across Poland and Scandinavia. These countries were showered with radioactive chemicals. The direct risks to health were inhalation and skin irradiation but there were other indirect risks. There were restrictions on the consumption of fresh vegetables and milk in several countries.

a Explain how radioactive materials get into (i) vegetables and (ii) milk. (I)
b Besides the potential hazards of accidents in nuclear power stations, state another problem which arises from generating electricity by nuclear means rather than by using tidal or solar energy. (I)
c One of the radioactive materials was iodine-131. Explain the link between this and its accumulation in the thyroid glands of people affected by the radioactive products. (I)

8 Describe how the cartoon illustrates three effects of Man on wildlife.

'So far, I've spotted thirteen hedgehogs, four rabbits, a squirrel, three rooks and a chaffinch.'

THEME 3 Organization and maintenance of the individual: structure and function

3.1 United we stand

Our framework – the skeleton

Your skeleton is the hard structure that supports your body (see Fig. 3.1.1). Your body is soft and flexible. It needs a skeleton to keep it from collapsing. The human skeleton consists almost entirely of bone. Most of

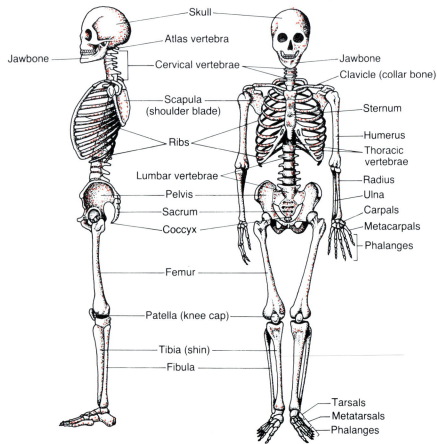

Fig. 3.1.1 The human skeleton: front and side views

the delicate parts of your body, like the brain and the heart, are protected by an armour of bone. The brain is surrounded by the skull. The spinal cord is enclosed within the backbone. The heart and the lungs are protected by the rib cage.

Your skeleton enables you to move your body. Nearly all the muscles of your body are attached to bone. The bony skeleton provides the rigid framework upon which the muscles can move the body. A muscle moves a bone by pulling it. Bones fit together at joints which are held fast by tough straps called **ligaments**. Some joints, such as the elbow and knee, are movable. Others, like those in the skull, cannot move.

The structure of the body is called its **anatomy**. The detailed anatomy of the skeleton has been known and understood for a very long time. After death, the soft parts of the body decay very quickly. The skeleton, however, takes a very long time to decay. So anatomists can study the structure of the skeleton at leisure. Skeletons of men and women who died several thousands of years ago have been found in very good condition.

There are 206 bones in the adult human skeleton. These range from the thigh bone at 45cm in length to the tiny bones of the ear which are less than 1cm long. Bone is a living tissue (see Fig. 3.1.2). It contains living cells, with a rich blood supply to feed them. The strength and hardness of bone is due to the crystals of calcium salts among the cells.

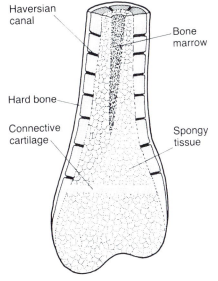

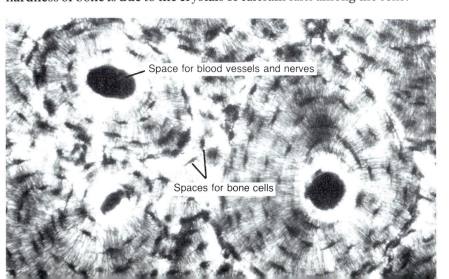

Fig. 3.1.2 (a) The inside of a bone; (b) microscopic structure of bone – the Haversian systems

Imagine that you need to make a number of strong cylinders, using a hard substance as the basic material. If you made the cylinders out of solid material they would be very heavy and you might quickly use up your material. Now if you made hollow cylinders, like pipes, they would be much lighter, less likely to snap, and contain less material. Many of your bones are made on this principle.

There are three basic types of bone construction in the body. There are long bones, short bones and flat bones. Long bones, like the bones of the arms and the legs, consist of hollow tubes. The shaft of a long bone is made of an outer tube of bone, called **compact bone**. In the middle of this tube is a tissue called the **bone marrow**. The swellings at the end of the bone are not hollow. They contain bone criss-crossed by a network of sponge-like bone tissue. It is called **spongy bone**. A short bone, like the bones of the hand, consists of a mass of spongy bone and marrow, surrounded by a thin layer of compact bone. Again this type of bone has the advantage of maximum strength with minimum weight. The third kind of bone – the **flat bone** – is found in the construction of the skull. Skull bones are relatively thin sheets of bone, but they need to be strong to protect the brain from injury. The flat bones of the skull consist of sandwiches of spongy bone between two strong layers of compact bone.

At the joints, where one bone moves against another, surfaces have to be smooth to ensure that there is little friction. In most joints there is a pad of a hard but smooth and flexible material called **cartilage**. The pads of cartilage between the bones of the back are often called **discs**.

The backbone, spine or **vertebral column** is the central support of the human skeleton. Connected to the top of the spine is the skull, which forms the bony foundation of the head. The shoulder, or **pectoral girdle**, is connected to the spine by a bone at the front of the chest called the **sternum**. The ribs are connected to the sternum. The bones of the shoulder can be moved freely in relation to the spine. Connected to the shoulder girdle are the bones of the arms. Towards the base of the backbone is the **pelvis** or **pelvic girdle**. The pelvic girdle connects the legs to the spine.

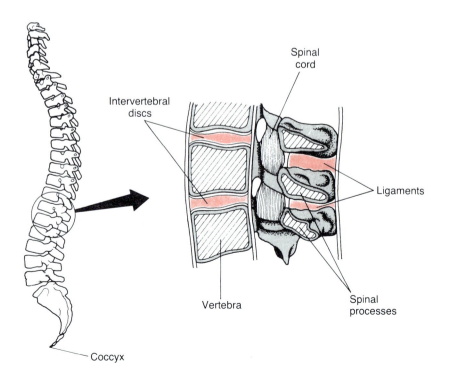

Fig. 3.1.3   The vertebral column

The spine or vertebral column of a young child is made of 33 separate bones called **vertebrae**. In an adult, some of these are fused together so that there are only 26 separate bones. The four vertebrae at the very end of the spine are joined together to form the **coccyx** (see Fig. 3.1.3). This is the vestigial tail or the remains of a tail. The coccyx is evidence that millions of years ago, Man's ancestors had tails (see p. 15). Tails became less and less useful as Man's ancestors developed into Man. Once our ancestors began to walk in an upright position, tails would have got in the way of movement. Thus, short tailed individuals were favoured by natural selection. Gradually, tailless species developed and we have probably evolved from these.

As the spine is made of so many bones, it is fairly flexible. The bones of the spine together form a hollow tube which encloses and protects the main nerve, the **spinal cord**. Each vertebra has a large central hole, through which the spinal cord runs (see Fig. 3.1.4). There are also spaces between the vertebrae, which allow nerves to go from the spinal cord to the surrounding tissues.

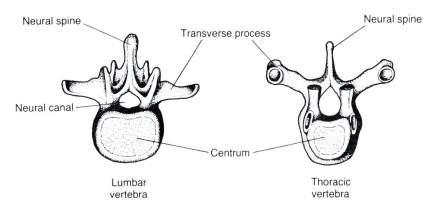

Fig. 3.1.4 Examples of vertebrae

Most of the vertebrae have bony projections, which provide bases for the attachment of muscles. These control the movement of the spine. Attached to each of the 12 chest vertebrae is a pair of narrow bones, called **ribs**. At the front, most of the ribs are attached to the breastbone, or sternum. The ribs, breastbone and part of the spine therefore form a protective cage around the delicate organs of the chest, the heart and the lungs. The two lowest pairs of ribs are only attached to the backbone, not the breastbone and so are called the '**floating ribs**'.

The skull is made of a number of bones. The top is called the **cranium**. It is made of eight bones joined by sutures. In the cranium of a baby, these bones are separate (see Fig. 3.1.5). As we grow older, they fuse together. There is a total of 22 bones in the skull.

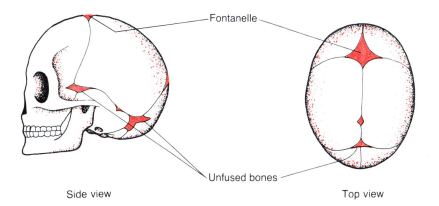

Fig. 3.1.5 The unfused bones of a baby's skull

Two collar bones or **clavicles** are attached to the top of the breastbone, one on each side. The clavicles are part of the pectoral girdle, which supports the arms. At the shoulder each collar bone is connected to a flat bone called the **scapula** or shoulder blade. The two shoulder blades lie flat on the back of the ribs. Each shoulder blade has a hollow into which the bone of the upper arm, the **humerus**, fits.

The joint of the upper arm and shoulder is called a '**ball-and-socket**' joint. The end of the upper arm bone is round and fits neatly into the socket in the shoulder blade. This kind of joint gives great freedom of movement. The joint at the elbow, however, is a '**hinge**' joint. This means that the joint can only move in one direction – a relatively restricted movement. The forearm consists of two bones, the **radius** and **ulna**. The radius can revolve around the ulna, allowing you to turn your hand over.

There is great freedom of movement at the wrist. The wrist is made up of eight short bones called **carpals**, arranged in two rows of four. These bones can be moved in any direction. The palm of the hand consists of five fairly long bones, called **metacarpals**. The bones of the fingers are called the **phalanges**. There are 14 phalanges in each hand – two in the thumb and three in each of the four fingers.

The pelvic or hip girdle is made up of three pairs of bones joined to make one structure. The pelvis is a hollow structure like a basin, in which many of the abdominal organs are contained. Each thigh bone, or femur, fits into a deep ball-and-socket joint in the pelvis. So like the shoulder joint, the hip is a very mobile joint.

The knee joint, like the elbow, is a hinge joint (see Fig. 3.1.6). At the knee, the **femur** connects with the larger bone of the lower leg, the **tibia**. There is a second leg bone on the outside of the tibia; this is the **fibula**. At the front of the knee joint there is a small bone called the kneecap, or **patella**.

The foot contains a total of 26 bones. There are seven short **tarsal** bones in the ankle. The five **metatarsal** bones of the foot correspond to the metacarpals of the hand. As in the fingers of the hand, there is a total of 14 toe bones, the phalanges.

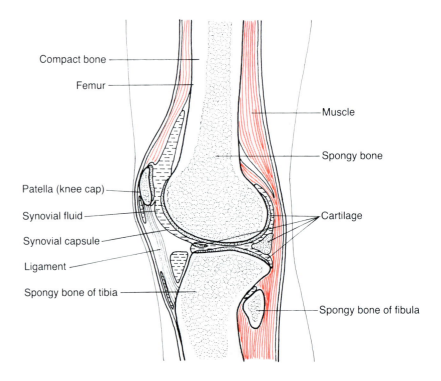

Fig. 3.1.6 A section through a knee joint

On the move   A skeleton without muscles is like a car chassis without a motor. Movements occur at joints when muscles are activated by nerves. When muscles contract across joints, they pull on bones to make them move. In this way the bones act like levers. Forces acting on a lever produce movements centred at a point, the fulcrum. Two forces are usually involved – the weight and the effort (pulling force).

Some levers are more efficient than others. They are grouped into first, second and third orders.

First order levers
The effort and the weight are on opposite sides of the fulcrum. The movement of the effort results in movement of the load in the opposite direction. When the fulcrum is central, the effort must equal the weight at equilibrium. This type of lever produces movements of the head and trunk with very little contraction of the muscles involved (see Fig. 3.1.7(a)).

Second order levers
The weight lies between the fulcrum and the effort. The movement of the effort results in movement of the weight in the same direction. The effort is less than the weight. This type of lever action is seen when you raise yourself up on tiptoe (see Fig. 3.1.7(b)).

Third order levers
The effort is exerted between the fulcrum and the weight. The movement of the effort results in movement of the weight in the same direction. The effort is greater than the weight. This is seen in the action of the arm and is the commonest type of lever in the body. A variety of movements can be made with very little shortening of the muscles involved (see Fig. 3.1.7(c)).

Fig. 3.1.7 The lever action of bones

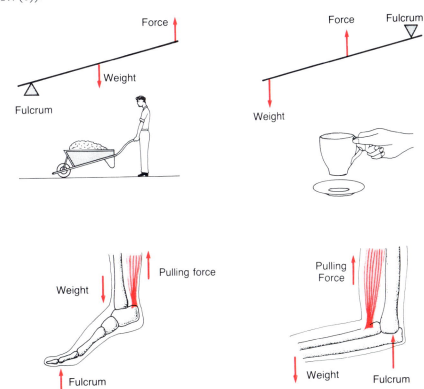

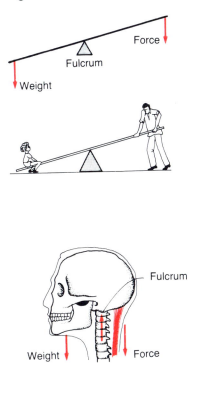

(a) FIRST ORDER LEVER   (b) SECOND ORDER LEVER   (c) THIRD ORDER LEVER

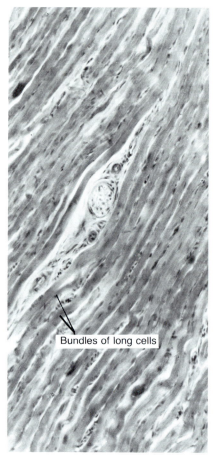

Fig. 3.1.8   Skeletal muscle –
photomicrograph of striated muscle

The muscles that cause the skeleton to move are made of bundles of long cells (see Fig. 3.1.8). Each cell contracts when a muscle works. The resting muscle cell is long and thin. In action it becomes short and fat. Muscles which move bones at joints work in pairs. One of the pair *contracts* while its partner *relaxes*. These are called **antagonistic pairs**.

Muscle attachment to bones is by means of tough, non-elastic **tendons**. In order to function there must be two firm points of attachment (see Fig. 3.1.9). The point on the bone which does not move when the muscle contracts is the **point of origin**. The point on the bone which moves is the **point of insertion**.

An active muscle needs oxygen and glucose as a fuel to supply energy. The release of energy produces carbon dioxide that has to be removed. So muscles need a good blood supply. Blood is supplied to muscles by **capillaries**. These tiny tubes open when the muscle is very active and close when it is resting. Muscles contract when they receive messages along nerves. The messages travel at more than 200 miles per hour.

Muscle cells contain a chemical called **adenosine triphosphate** (ATP). This can be called the energy currency of the cell as it has a rich store of energy within its molecules. The energy is used for all the body's needs. If there is a rich supply of oxygen to the muscle cells, then most of the energy in ATP is made available for contraction. However, if necessary, muscles can contract without oxygen. Under these conditions, a waste product called **lactic acid** builds up. With too much of this chemical, muscles cannot respond. Pain is often felt and a state of cramp can occur. Some of the lactic acid passes into the blood and warns the breathing control centre of the brain that more oxygen is needed. With the necessary oxygen, more lactic acid is oxidized to release energy.

In a resting muscle or in one doing moderate work, the oxygen is adequate. In violent exercise the muscle 'goes into debt' for oxygen. Lactic acid accumulates, to be removed when the oxygen supply allows. An athlete who has exerted himself to his limit has a large 'oxygen debt' in his muscles. This is 'repaid' with oxygen obtained in deep panting breaths of exhaustion.

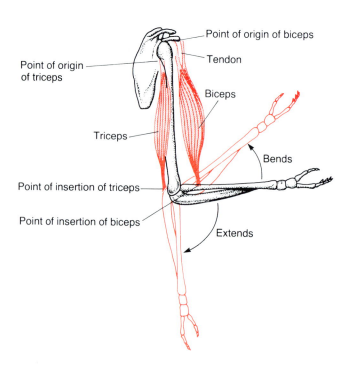

Fig. 3.1.9   Points of muscle attachment

Questions: Domains I and II

1 Complete the table by placing one tick in each column to indicate the type of joint present at each position in the body. (I)

Type of joint	Position in the body					
	Skull	Hip	Finger	Vertebral column	Shoulder	Knee
Ball and socket						
Fused (fixed)						
Hinged						

2 A man is seated at his desk writing when the telephone rings. Name three different types of joint where movement takes place as he puts down his pen, looks up and stretches forward across the desk to reach the telephone. In each case state the position of the type of joint named.

Type of joint	Position
1.	
2.	
3.	

b State the different types of muscle tissue used to bring about these movements. (I)

c Explain how the pairs of muscles act to bring about movement at a joint. (I)

3 Complete the table to give three functions of the skeleton and the bones or parts of the skeleton that carry out each function. (I)

Function	Bone or part
1.	
2.	
3.	

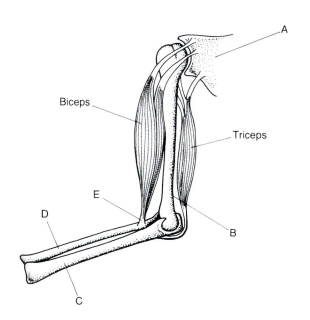

4

a Name the structures labelled A–D in the diagram. (I)

b Name the structure labelled E and give one reason why it is important for it to be non-elastic. (I)

c Name the substance which covers the ends of bones forming moveable joints. (I)

d When the arm is in a fully straightened position, name the part that can bring it back into the bent position shown. (I)

e From the list in brackets, choose the correct words to fill the spaces in the statements i to vi below. (I)
(attachment, pull, backbone, triceps, antagonistic, origin, lymphatic fluid, muscle tone, tiredness, push, pelvis, destination, synovial fluid, complimentary, insertion).

i The part of the skeleton which equally distributes the body's weight to the legs is the _____.

ii In a moveable joint, friction between the bones is reduced by _____.

iii The voluntary muscles are arranged in _____ pairs.

iv The position where the biceps muscle is joined to bone D on the diagram is called the muscle's _____.

v The position where the biceps muscle is joined to bone A on the diagram is called the muscle's _____.

vi Good posture is maintained by the regular contraction of individual muscle fibres working in turns to avoid fatigue; this is referred to as _____.

5 The skull is composed of two layers of solid bone, separated by softer spongy bone in the form of a lattice.

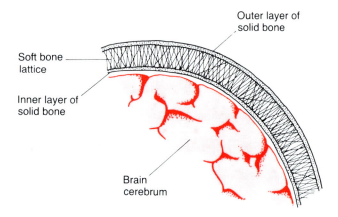

Outer layer of solid bone

Soft bone lattice

Inner layer of solid bone

Brain cerebrum

Give two reasons why this arrangement is more to our advantage than would be a skull of solid bone equal in thickness to these three layers. (I)

6 The table shows the units of lactic acid produced with time in the leg muscles of an athlete.

Time (minutes)	0	10	20	30	40	50	60	70	80
Units of lactic acid	0	1	7	12	9	6	3	1	1

a Plot a line graph from the data. (II)
b After what time did the amount of lactic acid reach a maximum? (II)
c During which 10-minute period did the lactic acid increase most? (II)
d What was the lactic acid level at
  i 25 minutes?
  ii 55 minutes? (II)

7 The diagram shows a comparison of time courses for contraction and relaxation of the three types of muscle: skeletal, cardiac and smooth.

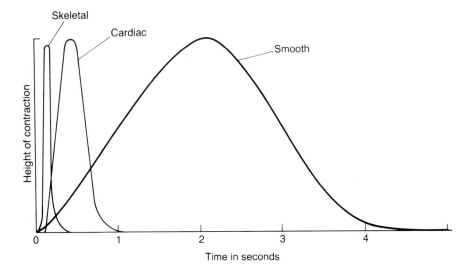

Skeletal

Cardiac

Smooth

Height of contraction

Time in seconds

a Suggest one reason why skeletal muscle should have the ability to contract and relax quickly. (I)
b Suggest one advantage of the slow contraction and relaxation of smooth muscle. (I)
c Name one organ in which smooth muscle may be found. (I)
d Suggest one reason why it is important for cardiac muscle to contract and relax more slowly than skeletal muscle. (I)
e State one way in which the cells of cardiac muscle are more complex than either of the other two muscle types. (I)

Experimental skills and investigations: Domain III

1 A practical demonstration of the action of the biceps and triceps

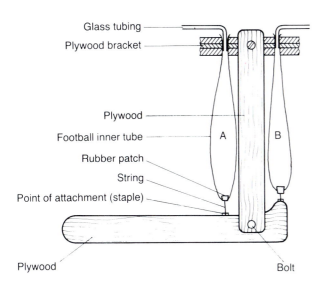

Glass tubing
Plywood bracket
Plywood
Football inner tube
Rubber patch
String
Point of attachment (staple)
A   B
Plywood
Bolt

Procedure
1 Set up the model as shown in the diagram. Make sure that the string is firmly attached to the rubber patch (use 'Superglue', taking great care). Use staples to pin the string to the hardboard.
2 Pump up the football inner tube with a bicycle pump.
3 Allow one inner tube to deflate while the other one is inflated.
a Describe the effect of inflating A and deflating B.
b Compare the observations with those on the movement of an arm.

2 How strong is your arm?

Procedure
1 Place your arm on a flat surface as shown opposite and get someone to hold your upper arm in position X on the diagram.
2 Fix a newtonmeter to a point which cannot move, below the bench as shown.
3 Without moving the upper arm, pull on the newtonmeter and record the maximum force you can exert.
4 After resting the arm, repeat the procedure with the upper arm held in positions Y and Z.
a Which is the best position to hold your arm when you lift things?
b Explain your answer.

3 Where does the energy come from for muscular contraction?

Procedure
1 Place some fresh meat (shin beef) in Ringer's solution and leave it at 35–40°C for 15 minutes.
2 Use mounted needles to tease out three or four muscle fibres from the meat about 20mm long.
3 Place the fibres in a drop of Ringer's solution on a microscope slide and place the slide on some graph paper. Measure the exact length (0.5mm) of the fibres.
4 Remove the excess Ringer's solution with filter paper, add a few drops of dilute glucose solution and leave for five minutes.
5 Measure the length of the fibres again.
6 Take a clean microscope slide and fresh muscle fibres and repeat stages 1–5 using a solution of ATP.

Use your observations to suggest where the energy comes from for muscle contraction. Explain your conclusion.

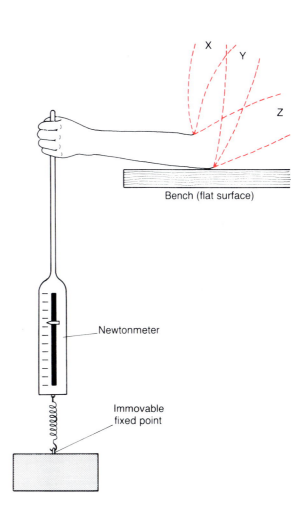

X   Y

Z

Bench (flat surface)

Newtonmeter

Immovable fixed point

3.2 Our daily bread

Fig. 3.2.1 'Our daily bread'

Our fuel and its uses

Fig. 3.2.2 A day's food for four people: (a) a business man – 2500 cals; (b) a housewife – 2650 cals; (c) a typist – 3600 cals; (d) a labourer – 4900 cals

Each person in a normal lifetime eats about 30 tons of food (see Fig. 3.2.2). Every year an increasing number of foods are being developed and modified to meet our needs. Early Man was a hunter and gatherer. He was dependent upon his skill in the chase. His knowledge of the whereabouts of nuts, fruits, roots and leaves was essential for an adequate food supply. Certain communities, even today, retain these methods of

Fig. 3.2.3 Man the hunter

obtaining food (see Fig. 3.2.3). It can be satisfactory as long as there is a plentiful amount of game to hunt throughout the year.

While men hunt, the women usually gather vegetable products. Frequently, food shortages arise. Disputes between hunting communities are not uncommon. A wide range of animal species is considered suitable for food. Large mammals, because they provide large quantities of meat, are most desirable. Other more easily trapped animals such as lizards and insects have also formed an important part of people's diet.

The cultivation of cereal grasses and the domestication of animals provided the foundation for a more complex society. The development of agriculture enabled a larger population to be supported (see Fig. 3.2.4). The cultivation of cereals was most productive in terms of food value per acre. Hunting communities find it difficult to make provision for future needs. Cereals, however, can be stored for months. The introduction of cereals as a major part of Man's diet meant that carbohydrates (see p. 112) provided a large part of his energy needs. Domestication of animals introduced a further group of foods – milk and milk products. From early times our food supply has been derived from several sources. These are cereals, meat, fish, eggs, dairy produce, sugar, fruits and vegetables. After many thousands of years our diet can still be divided into these categories. Few completely new sources of food have become available.

Fig. 3.2.4 Modern methods of agriculture: (a) harvesting cereals; (b) dairy farming

(a)

(b)

With Man's conquest of new regions, diet became characteristic of a particular area of the globe. Climate and soil favoured some crops in tropical regions and others in temperate zones. As a result, people tended to become dependent on a single food source to provide most of their energy. Thus rice became the staple food for South-East Asia (see Fig. 3.2.5), maize for Central America and wheat for Europe. There are dangers associated with the consumption of a single food as the only article of diet (see Fig. 3.2.6). No single food contains all the nutrients in their correct proportions to satisfy our needs. Deficiency diseases are associated with particular staple foods. For example, beri-beri is common where rice is the staple diet, pellagra with maize, and kwashiorkor with plantain and cassava diets (see Fig. 3.2.7).

Many changes in eating habits occurred in Britain and the USA during the industrial revolution of the nineteenth century. Industrialization created a large population dependent on a food industry. Farmers, transport workers and shopkeepers struggled to provide an adequate food supply. Urbanization in Britain and the USA at first resulted in a poorer diet for a large number of people. As material wealth developed the diet improved for all sections of the community. By the mid-twentieth century, serious dietary deficiency diseases were rare in these countries. African nations undergoing urbanization show similar problems. Suitable measures have not been taken to avoid the food shortages associated with rapidly growing populations. This has too often coincided with drought and famine.

Fig. 3.2.5 Rice became the staple food of South-East Asia

Fig. 3.2.7 Kwashiorkor

Fig. 3.2.6 There are dangers associated with the consumption of a single staple food like maize in Central America

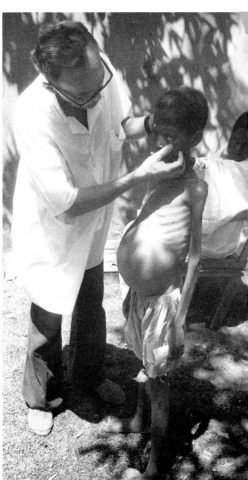

A study of food consumption in Britain and the USA in the last hundred years indicates changes in diet. The most noteworthy is the increase in sugar consumption and the decrease in cereal consumption. An increase in the proportion of energy provided by fat is also obvious. Meat and fruit consumption has also increased. The increase in 'convenience foods' has been explosive (see Fig. 3.2.8).

Increase in the consumption of sugar and fats is partly due to biotechnology. This makes possible the extraction of these and their addition to many food products. Convenience foods include packaged meals, frozen and canned meat, fruit and vegetables, cake mixes, breakfast cereals and packet soups. The increased consumption of these may lead to an increase in leisure time. They need little preparation and are popular with the modern housewife.

Fig. 3.2.8   Convenience foods

Meat and fruit are comparatively expensive foods. Economic factors tend to govern the level of consumption. Although the consumption of bread is declining, certain other cereal products are eaten in greater quantities. Similarly, liquid milk consumption in Britain is declining slowly. Increased consumption of cheese, yoghurt and cream offsets this to a certain extent. Availability also influences food consumption patterns. For example, beef shortages in Britain during the end of the 1960s resulted in high prices and a decrease in consumption. On the other hand, low-priced poultry meat consumption increased in the same period (see Fig. 3.2.9).

We eat because we feel hunger and attempt to relieve it. We select foods according to our appetites and preferences. The choice between foods may be made on the basis of palatability. We can distinguish and appreciate sweet, sour, bitter and salt tastes. We choose foods which provide an acceptable mixture. Tradition, culture, and availability influence the extent to which population groups prefer certain flavours. It is possible that nutritional needs lead us to desire certain flavours in food. Salt flavour generally leads us to meat; sweetness to fruit. These provide protein and vitamin C respectively.

Fig. 3.2.9 Battery-fed hens

Climate, soil and water govern which crops can be grown. In highly developed communities, transport and food technology are able to increase the availability of foods. For example, frozen and canned vegetables can be eaten throughout the year (see Fig. 3.2.10). Food technology also develops 'new' products by modifying food extracts and mixing them with other foods. Sugar, an extracted food, has at least two qualities which make it a good food additive. It is a preservative and its sweetness improves flavour. The addition of sugar to a great variety of foods is a common practice. Extracted fats enable us to use the properties of fat to 'improve' other foods. Fat provides a useful high-temperature cooking medium which increases the number of ways in which food can be prepared.

Fig. 3.2.10 Frozen and canned vegetables

Traditions and taboos   Many religions influence a person's choice of food. There may be strict laws as in the Jewish faith. Certain foods are prohibited altogether and methods of preparation are sometimes specified (see Figs 3.2.11 and 3.2.12). Vegetarianism may be a ritual and a discipline associated with religions. In some cultures pain or penalty is associated with the consumption of particular foods. In some societies eggs are excluded from the diet of women because they are believed to cause infertility. Cow's milk is rejected by others as food for children because it is believed to cause diarrhoea. Tradition, habit and custom also strongly influence food choice. Roast beef is eaten with Yorkshire pudding in England. Turkey is eaten at Thanksgiving in the USA. Similar traditions related to food exist in every culture.

Fig. 3.2.11   Indian pilgrim cooking atta, a type of flour preparation

Fig. 3.2.12   Typical Indian spice sellers

Fig. 3.2.13 Coca-Cola advertisement

Social factors may result in the increasing popularity of a particular food product. The success of Coca-Cola is world famous (see Fig. 3.2.13). It has become established as a permanent new feature in the diet of many people. Foods are frequently given special qualities and status. Modern Man's obsession with health has resulted in a search for foods that will achieve this end. Yoghurt, brown sugar, wholemeal bread, and honey all currently enjoy popularity. The desire in Western society for a slim figure has also given rise to 'slimming' foods (see Fig. 3.2.14). This anxiety about health is used by food manufacturers to sell their products. There is no doubt that advertising has a great influence on food selection.

Fig. 3.2.14 Slimming foods

The right balance   The normal diet needs to contain adequate amounts of protein, fat and carbohydrate, as well as certain minerals and vitamins. Apart from their specific functions in the body, the first three provide energy.

Protein   **Proteins** are complex compounds of nitrogen, carbon, oxygen, hydrogen and often sulphur and phosphorus. All living things depend on proteins for their growth (see Fig. 3.2.15). Proteins form parts of the body and many substances necessary for its proper working. The chemicals that pass on information from cell to cell when a person grows or reproduces contain proteins – the **nucleoproteins**.

Fig. 3.2.15   Foods rich in protein

Proteins are made up of molecules that are large when compared with those of other chemicals. (The **molecular weight** of a substance is the average weight of one of its molecules.) A protein may have a molecular weight of several thousand. That of water is only 18. Proteins in the blood are too big to pass through most membranes. However, they are important in the exchange of fluids and nutrients between blood and the cells.

Proteins are built up from smaller molecules, called **amino acids**. There are many hundreds of these in each molecule of protein. The amino acids are linked together to form a chain. Proteins differ from one another according to their amino acids and the shape the chains make. No one knows how many different proteins there are in the world. We do know that of the many types of amino acids only 23 have been found in proteins.

In the 'control centre' or **nucleus** of the cell is a chemical called **DNA** (**deoxyribose nucleic acid**). It contains a code for the order of amino acids in the protein chains. Different parts of DNA have the code for different proteins that are made in the cell. According to these instructions, amino acids are arranged in a particular order to make proteins. Amino acids not needed to make proteins are broken down further. The nitrogen part is removed from the body in urine. The remaining part of the amino acid can be used for energy or to make fat or carbohydrate.

If the necessary amino acids are not available, the body can change some amino acids that are not needed into ones that are. In humans there are eight that cannot be made in this way. They must enter the body in the diet. They are called **'essential'** amino acids.

If a person's diet lacks one or more of these essential amino acids, the working of the body may be disturbed. By adding a protein food that contains the missing amino acid the fault can be corrected. Egg protein has all the essential amino acids. Very few other foods contain all of them.

Young children who live on diets poor in protein suffer from a disease called kwashiorkor. They do not grow properly. Their hair loses its colour and becomes thin, and their tissues become swollen. The disease can be cured by adding protein to the diet. Kwashiorkor is common in Asia, Africa and South America. To be sure of a supply of all the essential amino acids, a variety of proteins should be eaten. An adult should have one gram of protein a day for every kilogram of body weight. A child should have about twice as much.

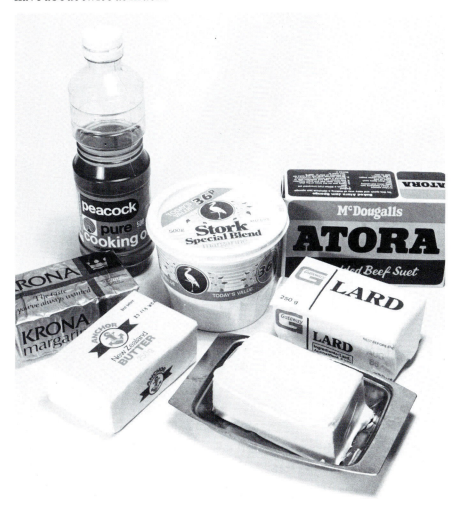

Fig. 3.2.16   Foods rich in fat

Fat   Part of our daily diet consists of **fat** (see Fig. 3.2.16). Butter, lard and oil are visible forms. Meat, eggs, milk and many others are invisible forms. Another name for fat is **lipid** and it provides us with a long term store of energy. Fats are large molecules composed of carbon, hydrogen and oxygen. These elements are arranged in such a way as to make a fat molecule of two basic parts. These parts are **glycerol** and **fatty acids**. Fat is broken down in the body's digestive system to these two components before it can be used by the body. Most authorities believe that animal (**saturated**) fats tend to increase the level of fats in the blood. Vegetable (**unsaturated**) fats do not increase the level of fats in the blood. In the diseases arteriosclerosis and coronary thrombosis deposits of fat form on the lining of arteries (see Figs 3.2.17 and 3.2.18). This blocks them and reduces the blood flow. The deposited fat is a type called **cholesterol**. Therefore anything that raises the blood cholesterol level could lead to coronary heart disease and heart attack (see p. 154). Anything that lowers it will help to avoid the condition. Cholesterol is found in some foods, particularly eggs and animals fats. As fat consumption has risen over the last hundred years, so has the occurrence of heart disease. Most people who have had a heart attack are encouraged to eat less animal fats. So why eat fat at all?

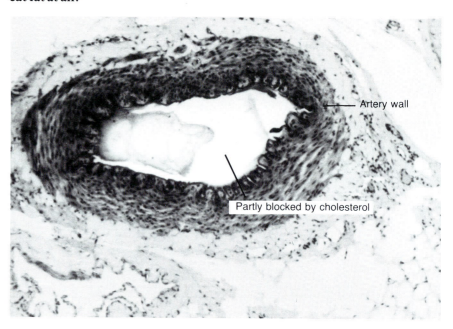

Fig. 3.2.17 Cross-section of an artery showing deposits of cholesterol

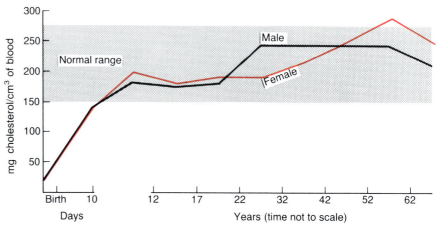

Fig. 3.2.18 Graph showing the variation of total blood cholesterol with age

First, fat provides energy. Weight for weight, it supplies more than twice as much energy as carbohydrate or protein. Most authorities recommend that in a balanced diet no more than forty per cent of the energy should come from fat. Half of the total should come from carbohydrate and ten per cent from protein.

Secondly, it is from fatty foods that we obtain the fat-soluble vitamins A and D. These vitamins are found in fatty foods.

Thirdly, fat makes a diet palatable. It is on this consideration that recommended intakes are based. With increasing affluence, fat consumption always rises. The level of about forty per cent seems to be what people like to have. That level is accepted as reasonable.

The total amount of fatty tissue varies widely from person to person. Normally, fat accounts for about 15 per cent of body weight. Most of this is in adipose tissue close to the skin. In obesity, the amount is greater (see Fig. 3.2.19). Although a heavy person will have increased amounts of water and protein, the greatest bulk of extra body-tissue is fat. Very obese people may have up to 70 per cent of their body weight as fat.

How the fat is distributed in a person of normal weight is fairly obvious to see. Men and women differ in this respect. Most fat stored in either sex is just below the skin. A standard method of determining a person's body fat is to pinch the skin with a pair of special calipers and see how much width is trapped between the two arms of the instrument. Towards middle-age, fat tends to be deposited around the abdomen (middle-age spread). All this fat under the skin acts to conserve heat in the body. In women more fat is laid down under the skin. The total fat content is higher. In particular, more is present on the hips and breasts. In some societies obesity has been considered essential to beauty. Apart from the direct effect of fat on cholesterol and hence on heart disease, obesity may lead to other diseases. Obesity may also be caused by eating too much carbohydrate. This can be changed into fat and stored in the body.

Fig. 3.2.19 Obesity

Carbohydrates What do a sugar lump, a loaf of bread and a potato have in common? They all contain carbohydrates (see Fig. 3.2.20). The sugar lump is made entirely of one of the simpler types of carbohydrate. The bread is made from flour, which contains a more complicated carbohydrate, **starch**. The potato also contains starch, food stored by the plant to feed its new shoots in spring.

Fig. 3.2.20   Foods rich in carbohydrate

There are many types of sugar. Plants make sugars by photosynthesis (see p. 29). Both plants and animals change these sugars into one type, **glucose**, which they use for energy. Plants store glucose by combining many molecules of it to form a large starch molecule. Starch does not dissolve in water and is too large to pass through cell membranes. When a plant needs the stored energy, it breaks down the large molecules into small glucose molecules and uses them. Glucose is a small molecule that dissolves in water and passes easily across cell membranes. Carbohydrates are the main source of energy in our diet. When we eat stored plant food in potatoes or flour, we break the starch down into glucose.

The name, **carbohydrate**, refers to the chemical make-up of these compounds. The formula shows that for each carbon atom (C) there are two hydrogen atoms (H) and one oxygen atom (O), as in water (H_2O). A sugar that contains six carbon atoms usually has enough hydrogen and oxygen atoms to make six molecules of water. Such a sugar would have the formula $C_6H_{12}O_6$.

Sugars are the simplest carbohydrates. The simplest sugars consist of just one type of sugar molecule. Such a molecule is called a **monosaccharide**, that is, 'one sugar'. Usually this molecule contains from three to seven carbon atoms. Ribose, a sugar that is found in the nucleus of the cell, has five carbon atoms. Glucose has six (see Fig. 3.2.21).

Two monosaccharide molecules can be joined together to make a double sugar, a **disaccharide**. We are most familiar with the disaccharide sucrose (see Fig. 3.2.21). This is the sugar we put in our tea. It is found in the sap of many plants.

The commercial sugar industry makes its brown or white crystals of sucrose from sugar cane and sugar beet. In warm climates these plants are fairly easy to grow in large quantities and yield plenty of sucrose. Honey also contains sucrose, mixed with another sugar, fructose (see Fig. 3.2.21).

Glucose

Sucrose

Fructose

Fig. 3.2.21 The chemicals present in sugars

Each molecule of sucrose is made up of a molecule of glucose and one of fructose joined together (see Fig. 3.2.21). These double molecules are too large to be absorbed by our cells. When we eat sucrose our digestive system breaks it down and changes it into molecules of glucose. All carbohydrates we eat are changed to glucose so that it can be used as a source of energy (see Fig. 3.2.22).

Every carbohydrate is made by plants. Green plants have tiny parts in many of their cells called **chloroplasts**. These contain the green pigment, **chlorophyll**. The chloroplast uses the energy in sunlight to make glucose. For raw materials, it uses carbon from carbon dioxide in the air, and hydrogen plus oxygen from water in the soil. The energy from sunlight binds these atoms together into glucose molecules. The glucose can then be used for energy or changed into other carbohydrates.

Fig. 3.2.22 The amount of energy used by the body in one hour for various activities

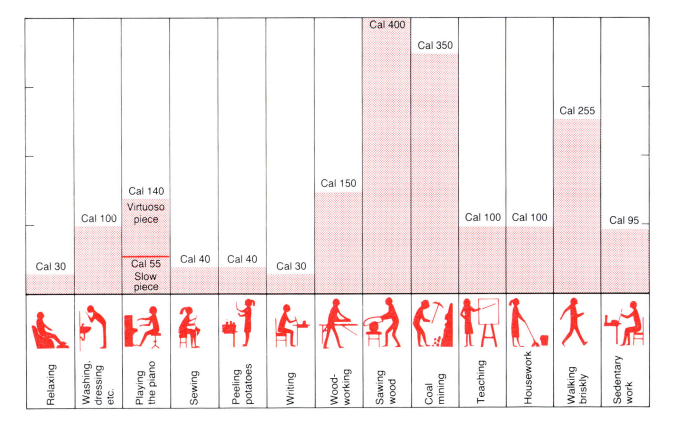

Relaxing	Washing, dressing etc.	Playing the piano	Sewing	Peeling potatoes	Writing	Wood-working	Sawing wood	Coal mining	Teaching	Housework	Walking briskly	Sedentary work
Cal 30	Cal 100	Cal 140 Virtuoso piece / Cal 55 Slow piece	Cal 40	Cal 40	Cal 30	Cal 150	Cal 400	Cal 350	Cal 100	Cal 100	Cal 255	Cal 95

When many sugar molecules join to form a large carbohydrate molecule it is called a **polysaccharide**. Glucose in plants is stored as the polysaccharide starch (see Fig. 3.2.23). Glucose in animals is stored as the polysaccharide **glycogen**. If we go without food for some time, some glycogen is turned into glucose to keep the level of glucose in the blood constant. **Cellulose** is a polysaccharide found in all plant cell walls. We cannot digest it and use it as the basis of '**roughage**' in our diet which aids digestion. We get most of our glucose by eating starch. The easiest way to eat it is in the storage organs of plants. Starch may be stored in underground swellings of the stem such as the potato. Tapioca is made from the swollen stem of the cassava plant. Storage roots include carrots and beet. Seeds like rice, wheat and corn contain a tiny embryo and a store of starch to feed it.

Starch forms 50 to 60 per cent of an average British diet. In poor countries it forms a greater part. Potatoes, rice, wheat, barley and corn are the basic part of most people's diet. At least two problems arise from eating too much carbohydrate. Extra glucose can be changed to fat causing problems related to obesity (see p. 111). The other problem is that bacteria which cause tooth decay also eat carbohydrates, especially sugar. If food debris is allowed to accumulate between the teeth, bacteria will act on carbohydrate and form acid, causing tooth decay.

Fig. 3.2.23   Three units of a molecule. Starch is made of hundreds of these units

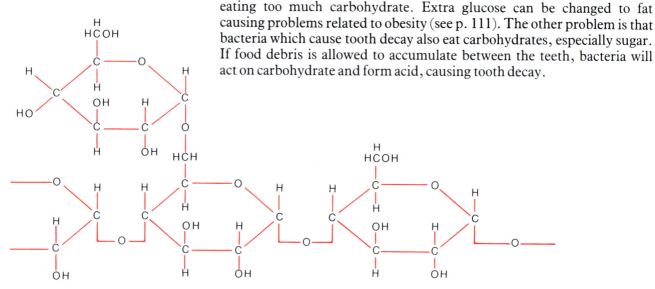

Summary of classes of food as percentage of a selection of foods

Animal/vegetable	% water	% protein	% fat	% carbohydrate
Fish (cod)	81	16	0.5	0
Egg	74	13	12	1
Meat	65	17	16	0
Soya beans	7	40	24	13
Sunflower seeds	5	30	40	25
Green beans	77	7	0.5	10
Peas	79	6	0	10
Milk	88	3.3	3.6	4.4
Rape seed	2	28	35	35
Groundnut	9	24	49	18
Wheat (bread)	37	9	1.2	51
Oats	9	12	9	66
Corn (maize)	12	9	4	74
Potatoes	78	2	0	16
Rice	12	7	1	80
Single cell (produced by biotechnology) (% on a moisture-free basis)				
Algae (*Spirulina maxima*)		62	3	17
Fungi (yeast)		52	2	30
(*Fusarium*)		57	1	41
Bacteria (*Bacillus pseudomonas*)		78	14	5

Vitamins and minerals

Vitamins are essential for the maintenance of good health (see Fig. 3.2.24). Many of them help enzymes as '**co-factors**'. Without them, the ezymes could not speed up chemical reactions in the body and symptoms of deficiency result. **Vitamin A**, obtained from milk, butter, eggs and vegetables, is essential for good eyesight. Any lack of this vitamin causes difficulty in seeing in dim light (night blindness). If the deficiency is severe, blindness can easily develop due to the cornea becoming opaque. Lack of vitamin A is one of the most common causes of blindness in the world today.

Vitamin B$_1$ is essential for the action of enzymes concerned with respiration in cells. Without the action of these enzymes, poisons such as pyruvic acid build up in the cells. It is found in meat and fish and in the outer coat of rice. When rice is polished in the preparation of white rice it is removed. If intake of B$_1$ is insufficient, beri-beri results with severe muscle weakness and heart disease.

Lack of **vitamin B$_2$** causes disease of the lips, tongue and skin. It is found in milk and meat. Another type of vitamin B is **nicotinic acid** or **niacin**. It is also found in meat and milk. Lack of it causes **pellagra**. It is the most common vitamin deficiency disease in the world today, particularly in the corn-eating countries. Symptoms include dermatitis, diarrhoea and breakdown of nervous tissue (see Fig. 3.2.25).

Vitamin C (**ascorbic acid**) is found especially in fresh fruit. If the diet is low in such foods, scurvy results (see Fig. 3.2.26). It shows itself by bleeding of the gums and skin.

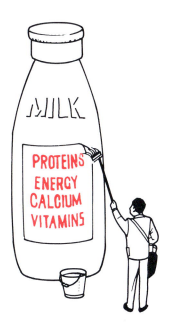

Fig. 3.2.24 During the Second World War, people were encouraged to get Vitamin A from carrots by a character called Dr. Carrot. Milk is also a good source of vitamins and minerals

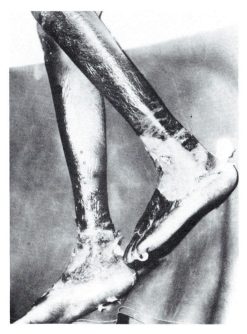

Fig. 3.2.25 The vitamin deficiency disease – pellagra

Fig. 3.2.26 Scurvy flew through the schooner's crew
As they sailed on the Arctic Sea.
They were far from land and their food was canned,
So they got no Vitamin C.
They were fed with pickled pork,
Those sailors bold and free,
Yet life's but brief on the best corned beef,
If you don't get Vitamin C.
(*from* St. Bartholomew's Hospital Journal)

Summary of the main food sources and functions for vitamins and mineral salts

Vitamin	Food sources	Functions	Effects of shortage
Vitamin A (this can appear as **carotene** in yellow and red fruits and vegetables)	Fortified margarine, dairy products, milk, oil (particularly cod-liver oil and halibut-liver oil), carrots, green vegetables	Promotes good eyesight, healthy skin and linings of the body, normal growth.	Skin disorders, night blindness.
Vitamin B Group **B₁ Thiamin**	Bread, flour, meat, potatoes, milk, fortified breakfast cereals	Helps to release energy from sugars and starches.	Lack of energy. Severe shortage causes **beri-beri**.
B₂ Riboflavin	Milk, meat (particularly liver), eggs, yeast extract	Helps to release energy from food, especially amino acids and fat. Promotes normal growth in children.	Severe shortage is rare, but when it does occur the lips, tongue and skin are affected.
Nicotinic acid	Meat, bread, flour, fortified breakfast cereals, milk, liver, kidney, yeast extract	Helps to release energy from food, especially carbohydrate. Essential for normal growth.	Nervous and digestive systems are affected. Severe shortage causes **pellagra**.
B₁₂ Cobalamin	Produced in the intestines by bacteria, so it mainly comes from animal foods. Liver is the richest source	Helps the metabolism of amino acids and other body enzyme systems.	Special types of anaemia can occur where red blood cells become enlarged. Vegans are at risk because they do not eat any food from animals.
Folic acid	Offal, raw green leafy vegetables, pulses, bread, oranges, bananas	Helps B₁₂ to divide cells.	Anaemia. Elderly people, pregnant women and premature infants could be at risk if there is a shortage in their diet.
Vitamin C (sometimes called **ascorbic acid**)	Green vegetables, citrus fruits (lemons, oranges, limes) black-currants, rose-hip syrup	Provides healthy connective tissues. Helps to rebuild new tissues and heal wounds.	Bleeding, particularly of the gums. Slow healing of wounds. Severe shortage causes **scurvy**.
Vitamin D	Fortified margarine, fatty fish, eggs, butter, and the action of sunlight on the skin	Maintains the level of calcium in the blood to form strong bones and teeth.	**Rickets** in babies and young children. Bone softening (**osteomalacia**) in adults. Possible tooth decay.

Mineral	Food sources	Functions
Calcium	Milk, cheese, fortified flour, green vegetables, bones in canned sardines and salmon	Forms the structure of bones and teeth. Helps to control the formation of muscles.
Sodium	Common salt, seafood, vegetables	These two together control the amount of water in the body.
Potassium	Meat, fish, fruit, vegetables	
Iron	Meat, bread, flour, other cereal products, potatoes, vegetables	Helps to form **haemoglobin** in red blood cells which carries oxygen needed for energy production.
Iodine	Tap water, seafood, vegetables, iodized salt	Small amounts needed to produce **thyroxine** which helps the thyroid gland work properly.
Fluorine	Fluoridated tap water, seafood, tea	Helps to form strong enamel on teeth.

The **mineral** element **calcium** is needed to build bones and teeth. It can only be absorbed from our food if there is enough **vitamin D**. This vitamin is found in eggs, fish and cod liver oil. It can be made in the skin under the action of sunlight. This explains why the deficiency disease rickets is often seen in countries where sunlight is poor. This is sometimes true of industrial cities where ultraviolet light is blocked out by pollution.

We can also suffer if our diet contains too little of the **trace elements**. Some of these have been extensively studied – iron, iodine, fluorine and cobalt, for example. Others are only now being researched in depth, for example, copper, zinc, chromium and selenium. Everyone needs **iron** for the manufacture of red blood cells and the prevention of anaemia (see p. 145). Women need more iron than men because of blood loss during menstruation. Pregnant women also need more iron than the average man.

Iodine is needed for normal functioning of the thyroid gland (see p. 189). Children whose diet is lacking in it cannot produce the growth hormone **thyroxin,** which controls physical and mental development. Iodine is often added to salt to overcome any deficiency in the diet. **Fluorine** is needed for healthy teeth while **cobalt**, in **vitamin B$_{12}$**, is needed for the manufacture of red blood cells.

Of the other minerals, it is thought that **zinc** deficiency might cause stunting of growth, as seen in Middle Eastern countries.

Too little too late

Malnutrition and starvation

Malnutrition results from a diet which is inadequate. It can be due to a lack of food or else the diet can contain too much of the wrong types of food. The degree of malnutrition can vary. Mild cases occur in the elderly poor in modern Western countries. Extreme forms result in starvation and are found most frequently in the developing countries of the Third World (see Fig. 3.2.27).

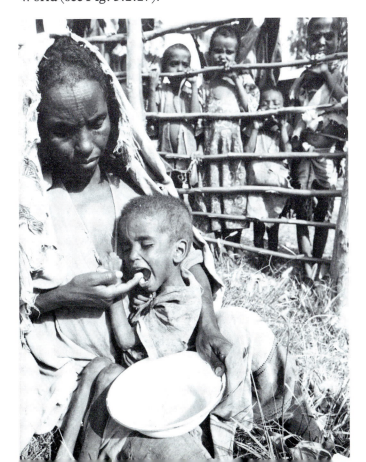

Fig. 3.2.27 Starvation

Mankind has always known famine. When crops fail vast numbers of people starve to death. The usual causes of famine are lack of rainfall, floods or earthquakes. Destruction of the harvest by pests such as locusts or crop diseases also occurs. An example of the widespread misery caused by crop disease was the Irish potato famine in the last century. When the crops were destroyed by a fungus, millions of people starved (see Fig. 3.2.28). Potatoes were the major part of their diet at that time.

Malnutrition is not solely due to natural disasters. Man's inhumanity to Man has been responsible throughout history for famine and starvation. There are many examples of the burning of the enemy's crops during wartime.

Deliberate under-feeding of prisoners in camps or ghettoes has often occurred. Malnutrition is not just a problem of the past. Inadequate supplies, and lack of facilities to distribute what is available, still contribute to a world-wide problem. Only about half of the world's population is properly nourished. The problem becomes even worse when natural disasters or war occur.

The world does not grow enough food to feed the population. This is a situation that the United Nations, through its Food and Agriculture Organization (FAO), is attempting to remedy. It is made worse by the ever-increasing number of mouths to feed. The population continues to rise alarmingly in many developing countries. This is mainly because of the drop in infant mortality. Efforts to promote birth control are meeting with only limited success. Some predict famine on a scale never seen before if the present situation is allowed to continue.

Groups at risk   Malnutrition also occurs in Western countries including the USA. Family doctors often treat patients suffering from lack of essential food materials. There are certain high-risk groups, for example, children, pregnant women and the elderly (see Fig. 3.2.29).

Children need food for their normal growth and development. This is particularly true during the first few years of life and again at puberty. Although not as common as it was, **rickets** is a serious disease caused by lack of vitamin D. A softening of the bones leads to stunting of the child's growth. This vitamin is now added to various foods for example, margarine, evaporated milk and baby foods. This had led to an almost complete elimination of the disease in developed countries.

Fig. 3.2.28   The Irish potato famine of 1847

Pregnant women need a greater supply of the right foods. They must provide for themselves and for the developing child. Lack of iron and vitamin B_{12} is seen frequently in pregnant women, causing severe anaemia. These substances are now added to the diet through foods available in most ante-natal clinics. Mothers who breast feed also need to protect themselves against the loss of nutrients in the milk. In extreme cases of malnutrition, the composition of the milk itself is poorer than normal.

The elderly members of the community are another group that frequently suffer from malnutrition. The inability of many old people to afford to buy essential foods is a major influence. The elderly who live alone are particularly at risk. They often are disinclined to prepare adequate meals. Scurvy is still quite common in this group. Badly fitting dentures or lack of them also contributes to malnutrition in old people.

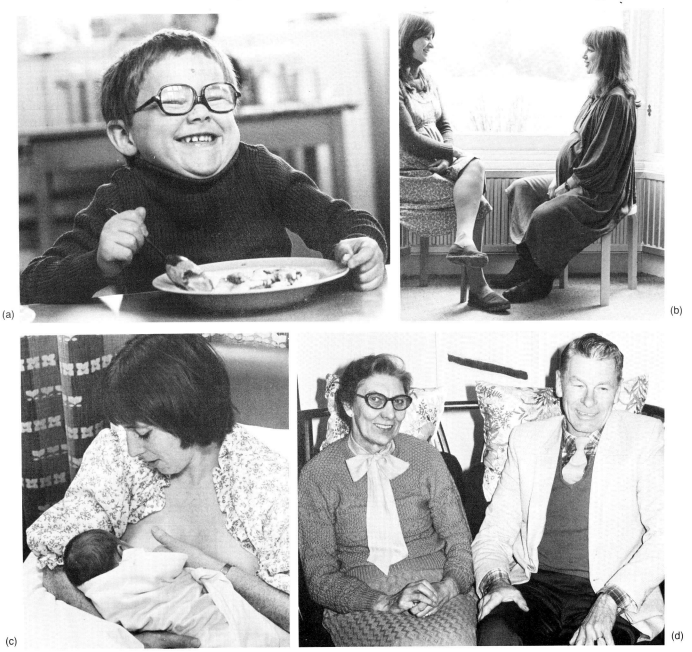

(a)

(b)

(c)

(d)

Fig. 3.2.29 High-risk groups for malnutrition (a) Children; (b) Pregnant women; (c) Mothers breastfeeding; (d) The elderly

Most people who are ill are disinclined to eat. Usually, if the illness is treated, the appetite returns. Sufferers from prolonged illness remain permanently uninterested in eating. They present the additional symptoms of malnutrition. Alcoholism also results in malnutrition. Severe alcoholics drink almost continuously. When hunger forces them to eat they usually buy the cheapest food. Their diet is often low in proteins and vitamins.

Malnutrition can also be seen in people trying to lose weight. Those on very strict diets without proper advice often do themselves harm. Beneficial foods are often omitted from the diet. Very strict vegetarians have to obtain their protein from a carefully planned diet of vegetable foods.

As a community becomes more prosperous the risk of malnutrition generally becomes less. However, social and economic factors may contribute to malnutrition. One such factor is the use of processed or synthetic foods. These frequently lack the vitamins and minerals found in natural foods. This is particularly true of baby foods. Many countries have introduced laws to control the food value of these products. Food preparation can destroy the health-giving constituents of some foods. Vitamin C is easily destroyed in cooking.

The effects of malnutrition may be seen only on close examination of a patient. Gross wasting is obvious in a person suffering from starvation. One of the most common signs of malnutrition is loss of weight. It may be a gradual process, only noticeable on regularly weighing a patient. Another striking feature of severe malnutrition is the amount of fluid kept in the body. The legs and abdomen become swollen. The swelling is mainly due to lack of protein. When the protein level falls water oozes out of the blood into the body tissues.

When all the essential foods are lacking, the malnutrition is called **marasmus** (see Fig. 3.2.30). **Kwashiorkor** results from diets which are low in protein but contain many of the other essential foods. It was first described in 1931 by Dr Cicely Williams in Ghana. Many of the victims suffered severe skin diseases and muscle wasting. There is much fluid retention. It was found that the symptoms usually appeared when a child finished feeding from the mother's breast. Breast milk was adequate to supply most of the child's dietary needs. After weaning the diet lacked protein. The energy requirements were supplied mostly by carbohydrates.

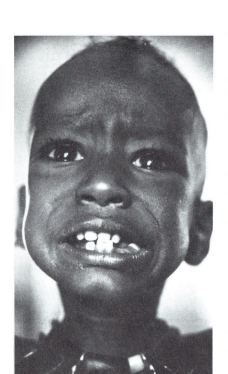

Fig. 3.2.30   Marasmus

Prophets of doom

How is malnutrition avoided? The difficulty is knowing (1) which foods to give; (2) in what amounts and (3) how to get the greatest quantities of the right food where it can do the greatest good for the greatest number of people.

In the highly developed countries high-risk groups can be helped. Hence, pregnant women are given more milk, iron and vitamins. Initially the needs of the new-born baby are met by the mother's breast milk. The only important additions to the diet are the vitamin C supplied as fruit-juice and vitamins A and D supplied in cod-liver-oil or a similar product. Both breast milk and cow's milk are deficient in iron. If the mother has adequate amounts of correct foods during pregnancy, her baby should have enough iron stored in the body to last for about four months. After this age the baby should be given foods rich in iron. During the early months there is also enough protein in milk to meet the baby's needs. When the baby is weaned off the breast it is necessary to supplement the diet with increased protein.

The elderly can be helped in many ways. One obvious approach is to increase their pensions. This would enable them to buy foods rich in protein and vitamins. The 'meals-on-wheels' service in some countries helps to ensure that the elderly have a well-balanced diet.

The problems in prosperous countries are small compared with those faced by developing countries. One failed harvest can bring famine and human misery on a scale almost beyond belief. Speed in the transportation of food is essential. Any delay in the shipping or flying out of vital supplies results in greater numbers starving to death (see Fig. 3.2.31). Unfortunately, international efforts are rarely coordinated. Much time and food is wasted by political manoeuvring and unscrupulous people out for individual gain.

The United Nations organization formed the Food and Agriculture Organization (FAO). Its aim was to look into ways of improving the food yield of any country that asked for advice. The problems faced by the FAO fall into two main groups: (1) how to improve the quantities of food produced and (2) how to supply new foods rich in vital nutrients. Some communities live solely on a single staple diet and suffer deficiencies as a result. High priority is given to ways of increasing amounts of food rich in protein. So often, crops that are easiest to grow have a low protein content. Cereals are a prime example. These consist mainly of starch. In many countries meat and fish are scarce and so the population lacks protein. The long-term answer is to promote the rearing of animals for food (see Fig. 3.2.32). In the short term, protein-rich foods can be provided in the form of fish meal or soya. Oil seeds, such as groundnuts or oil seed rape, have a relatively high protein content. On a smaller scale, fish farming may have a future together with the production of protein from micro-organisms. Some foods, although deficient in protein, are still valuable since they provide other necessary food factors. Potatoes, yams and cassava provide many of the B vitamins. Fruits and vegetables have a high vitamin C content. Help with irrigation schemes, pest control and the provision of education by prosperous countries could help produce food for underprivilaged nations.

Man is finding it impossible to feed the population of the world in the 1980s. What are the prospects for the year 2000? There are many scientists who predict world-wide shortages of food in the next generation. Others take a more optimistic outlook. They see the problem being solved by higher food yields and better land use. An increasing use of synthetic foods and a greater use of sea-farming are seen as other possibilities. The situation is unlikely to improve if the population explosion continues. By the year 2000 the world's population could be seven thousand million. The problem is made worse because the highest growth rates are in the developing countries. It is a vicious circle which must be broken. The world must provide more food or fewer mouths, preferably both.

Fig. 3.2.31   Bob Geldof's Band Aid appeal attempted to bypass the 'political' obstacles to food distribution

Fig. 3.2.32   Animal husbandry in the Third World

Questions: Domains I and II

1 The table below shows the approximate number of kilocalories and amount of protein needed by different people each day.

	Age (years)	Kilocalories	Protein (g)
Males	13–15	3200	85
	16–20	3800	100
	25	3200	65
	45	2900	65
Females	13–15	2500	80
	16–20	2400	75
	25	2300	55
	45	2100	55

a From the data deduce the following:
   i The sex which requires most energy.
   ii The age range of males which requires most energy.
   iii The age range of females which requires most energy.
   iv The difference in mass of protein needed by a male compared with a female at 25 years of age. (II)
b A boy aged 18 years, living in a developing country, eats 300g of protein per week. How much more protein is needed to make up the deficiency? (II)

2 The table shows the vitamin C content (mg per 100g) of certain foods.

Apples	5	Cauliflower	64
Blackcurrants	200	Grapefruit	40
Brussels sprouts	87	Lettuce	15
Cabbage (raw)	60	Potatoes (in December)	15
Cabbage (cooked)	18	Tomatoes	20

a Which of the above foods contains least vitamin C? (II)
b The recommended daily intake of vitamin C for a person 15 to 17 years old, is 30mg. What mass of raw cabbage would satisfy this requirement? (II)
c Give a reason why cooked cabbage has less vitamin C than raw cabbage. (I)

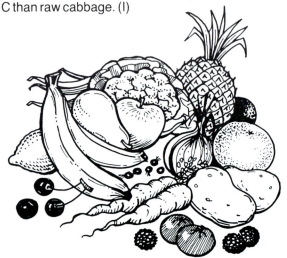

3 The pie charts show the proportions of water, carbohydrates, proteins and fats in some common foods.

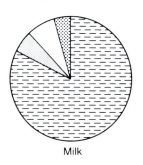
Milk

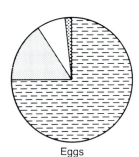
Eggs

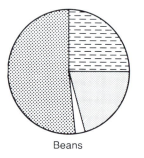
Beans

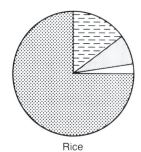
Rice

Key

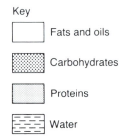

☐ Fats and oils

▦ Carbohydrates

☐ Proteins

▤ Water

a State   i   the food with most water;
        ii   the percentage of water in an egg;
        iii   the food with most carbohydrate;
        iv   the food with least protein. (II)
b If you were lacking in protein in your diet, which of the foods in the pie charts would you use to supplement your diet? (I)
c Name two other essential substances in your diet, not shown in the pie charts. (I)
d Give a reason why each of these substances is essential. (I)
e Which one of the four foods is the staple diet in many developing countries? (I)

Experimental skills and investigations: Domain III

1 Food tests: Carbohydrates

a Starch

Procedure
1 Put a suspension of food to be tested in a test tube.
2 Add a drop of iodine dissolved in potassium iodide.
3 Record your observation.

A blue-black colour indicates the presence of starch.

b Reducing sugar

Apparatus

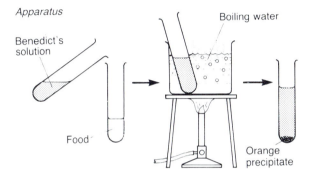

Procedure
1 Put $2cm^3$ of a solution of the food to be tested in a test tube.
2 Add $2cm^3$ of Benedict's solution to the test tube.
3 Place the test tube in a water bath with boiling water for two minutes.
4 Record any colour changes and other changes in the contents.

An orange precipitate (powder) indicates a reducing sugar.

c Non-reducing sugar

Apparatus

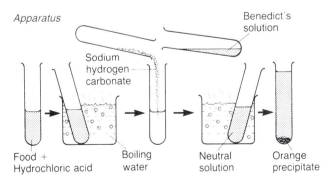

Procedure
1 Carry out a test for a reducing sugar as described above.
2 If no orange precipitate is formed, take $2cm^3$ of the food solution and add $1cm^3$ of dilute hydrochloric acid.
3 Place the test tube in a water bath with boiling water and leave it for two minutes.
4 Cool the test tube in running water and add sodium hydrogen carbonate until fizzing stops. The solution should now be neutral.
5 Add $2cm^3$ of Benedict's solution and place the tube in a water bath of boiling water for two minutes.
6 Record your observations.

Any non-reducing sugar that was present would have been changed to a reducing sugar by the hydrochloric acid. Therefore you can assume that a non-reducing sugar was originally present if the final test with Benedict's solution produced an orange precipitate.

2 Food tests: Fats

Apparatus

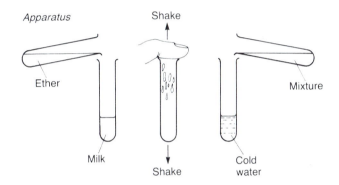

Procedure
1 Pour $2cm^3$ of milk into a test tube and add four drops of ether.
2 Shake the test tube with your thumb over the top.
3 Release the pressure built up in the test tube by removing your thumb from the top.
4 A clear layer should form on top of the milk. This is ether plus any fat present.
5 Pipette off the clear layer and drop it on to some filter paper.
6 Any fat will leave a greasy translucent (allows light through) mark.
7 Repeat stage 5 with a few drops of pure ether.
8 Record your results.

3 Food tests: Protein

Apparatus

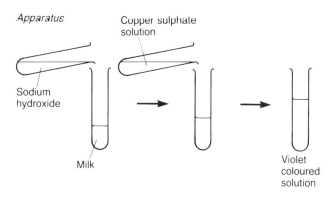

Copper sulphate solution

Sodium hydroxide

Milk

Violet coloured solution

Procedure
1 Place 2cm³ of a solution or suspension of the food to be tested in a test tube.
2 Add 1cm³ of dilute sodium hydroxide.
3 Add 1 per cent copper sulphate, carefully drop by drop, from a pipette.
4 Record your observation.

A violet colour indicates the presence of a protein.

4 A test for vitamin C

Procedure
1 Place 2cm³ of 0.1 per cent DCPIP (dichlorophenol indophenol) into a clean test tube. The observations in this experiment will depend on very precise measurements, so use a syringe to measure the exact volume.
2 Suck 2cm³ of the liquid to be tested (fruit juice) into another syringe.
3 Add the fruit juice, carefully drop by drop to the DCPIP and record how many drops it takes to change the colour of the DCPIP.
4 Repeat the procedure for a variety of liquids and find out which liquid contains the most vitamin C.

QUESTION
State the nature of the colour change that you observed when you added the fruit juice to the DCPIP.

5 A test for iodine

Procedure
1 Add coarse crystal sea salt (obtainable from a health store) to a test tube until it is about half full.
2 Add 3cm³ of 20 vol. hydrogen peroxide to the salt.
3 Shake the contents of the tube thoroughly.
4 Add a few drops of starch suspension, in distilled water, to the mixture.
5 Record your observations.

QUESTION
What was the purpose of the hydrogen peroxide?

6 How much energy is there in a peanut?

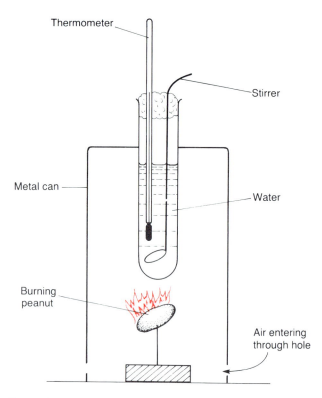

Thermometer

Stirrer

Metal can

Water

Burning peanut

Air entering through hole

Procedure
1 Weigh a large peanut, mount it as shown in the diagram and make sure that the tin can is well ventilated as indicated by the holes in the tin.
2 Place 20cm³ water in the test tube.
3 Note the temperature.
4 Ignite the peanut with a Bunsen burner but do not let the Bunsen burner heat the water in the test tube.
5 As soon as the peanut is burning, place the test tube over it to obtain as much heat as possible.
6 Note the temperature as soon as the peanut has burned itself out.
7 Calculate the temperature increase.

QUESTION
Given that 4.2 Joules are needed to raise 1g water 1°C and that 1cm³ of water weighs 1g, calculate the number of joules of heat that the water has received.
a Suggest why your answer is not exactly the same as the amount of energy that was in the peanut.
b Describe how you would alter the apparatus to give you a better idea of the amount of energy in the peanut.
c Why is it important to have holes in the tin as shown?

3.3 How food is put to use

Food is the only source of energy and of building materials in the body. Before food can be absorbed into the bloodstream it has to be broken down physically and chemically for cells to use. To do this the body has **teeth** and an **alimentary canal**. The alimentary canal is a hollow tube of about nine metres, stretching from mouth to anus. It is lined with special secretory cells and connected to nearby glands. Inside it, food is broken down chemically and then absorbed. The digestive system has to function without digesting itself in the process. It must move food along at the rate best suited to digestion and absorption.

By secreting biological catalysts called **enzymes** and **hormones**, it works with remarkable precision. **Digestion** is the breakdown of large insoluble molecules of food into small soluble molecules using digestive enzymes.

Mechanical breakdown of food – your teeth

There are four types of human teeth and each type works in a slightly different way (see Fig. 3.3.1).

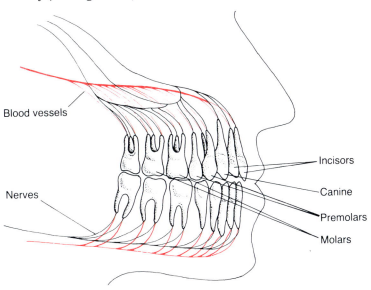

Blood vessels

Nerves

Incisors

Canine

Premolars

Molars

Fig. 3.3.1 The four types of teeth

1 **Incisors**
These are the teeth at the front of the mouth. Their purpose is to cut and bite food. Their edges are flat and sharp.

2 **Canines**
The visible part of these teeth, the crowns, have prominent points, called **cusps**. The teeth have large, deeply embedded roots. Their purpose is to tear food.

3 **Premolars**
These are partly used to tear, and partly to grind food. They have two cusps and one or two roots.

4 **Molars**
These have four or five cusps and are responsible for chewing and grinding food. The upper molars have three roots, the lower ones two.

In total there are 32 **permanent teeth**: 8 incisors, 4 canines, 8 premolars, and 12 molars (including 'wisdom teeth'). They replace the 20 deciduous or **milk teeth** which are: 8 incisors, 4 canines and 8 molars.

Structure of the tooth
(Fig. 3.3.2)

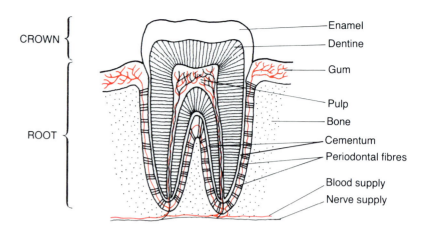

Fig. 3.3.2   The structure of a tooth

A tooth consists of two main sections:

1   The **crown**. This is above the gum and forms the biting surface.

2   The **root**. This is the part that is embedded in the jaw bone.

The parts of the teeth

1   **Enamel**
This is the outer covering of the crown and is the hardest substance in the body.

2   **Dentine**
This forms the bulk of the tooth. It is tough but not as hard or resistant to decay as enamel.

3   **Pulp**
This contains the nerve and the blood supply to the tooth.

4   **Cementum**
This is the hard rough covering of the root.

Teeth are firmly held in place and supported by the following:

1   **Bone**
The bone provides a socket for the tooth.

2   **Gum**
This protects the bone and roots of the tooth.

3   **Peridontal fibres**
These are thousands of tiny fibres which anchor the teeth.

4   **Nerves and blood supply**
The nerves carry sensation to the brain, whilst the blood vessels carry oxygen and nourishment for the teeth.

Chewing, swallowing and digestion

Digestion begins in the mouth. The incisor teeth cut food and the molar teeth grind it. In chewing, the molars exert a force which may exceed 500 kg per square centimetre. Thorough chewing exposes the largest possible surface area of food to the digestive enzymes. By softening the food, chewing eases its passage through to the stomach.

Saliva softens and lubricates food. It also begins carbohydrate digestion. It is produced at a rate of a litre a day from three pairs of **salivary glands**: 1 the **parotids** which lie just below and in front of the ears; 2 the **submaxillaries** and 3 the **sublinguals** located beneath the floor of the mouth. Salivary glands work as a result of seeing, tasting or

Fig. 3.3.3 The swallowing mechanism

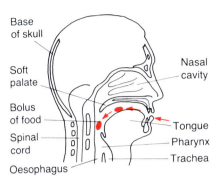

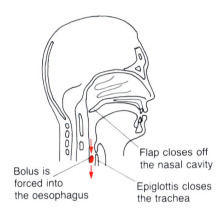

smelling food. These sensations cause a reflex action, known as mouth-watering. The largest flow of saliva is produced through an unconditioned reflex. It is triggered by the presence of food in the mouth. The saliva itself plays a part in tasting. It acts as a solvent for food flavours, increasing their effects on the taste buds. If a person likes what he tastes, he produces far more saliva than if he dislikes it.

Saliva is composed of two different types of secretion. The sublingual glands and submaxillary glands make a sticky mucous secretion called **mucin**. Mucin acts as a lubricant for food in the mouth and prepares it for swallowing. The parotid glands and submaxillary glands secrete an enzyme, called **salivary amylase**. In the slightly alkaline mouth it acts on the starch in foods, changing it into the two-unit sugar, maltose. Before food is swallowed, three to five per cent of starch is changed in this way.

Food is worked by the teeth and saliva. It is moulded into a soft mass, the **bolus**, by the tongue. Then it is moved backwards by the tongue to the back of the throat, the **pharynx** (see Fig. 3.3.3). The touch of the bolus on the wall of the pharynx begins the swallowing reflex. The wind-pipe or **trachea** is closed by a flap called the **epiglottis**. This prevents food from getting into the air passage leading to the lungs. Breathing is impossible at this stage.

Another flap closes off the openings to the inner nostrils. This prevents food passing out through the nose. Waves of muscular contraction force the bolus into the oesophagus (see Fig. 3.3.4). It takes between one and two seconds.

The **oesophagus** is 20 to 25 cm long. Food is moved through it to the stomach. The automatic movement of food along the gut is called **peristalsis**. It is made possible by layers of muscle present throughout the whole length of the gut wall. Peristalsis begins when the muscle contracts in a ring around the tract. This ring moves along, pushing food in front of it. It looks like fingers squeezing the contents of a tooth-paste tube. Food now enters the stomach through a ring of muscle called the **cardiac sphincter**.

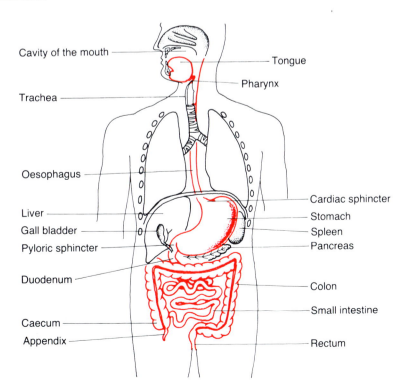

Fig. 3.3.4 The digestive system

Fig. 3.3.5   The position of the stomach after removal of the liver

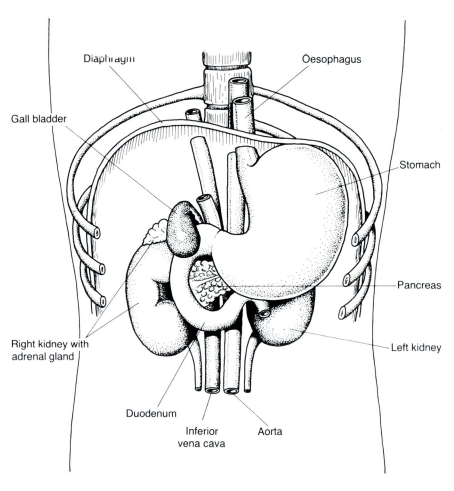

Diaphragm

Oesophagus

Gall bladder

Stomach

Pancreas

Right kidney with
adrenal gland

Left kidney

Duodenum

Inferior
vena cava

Aorta

The stomach   The **stomach** is a muscular bag about 30cm by 10cm (see Fig. 3.3.5). It has a volume of one and a quarter litres. It (1) stores food; (2) mixes it and (3) starts the digestion of protein. Apart from a small amount of alcohol and glucose uptake, no absorption occurs in the stomach. The inner surface is much folded. Embedded in it are two sorts of glands – **gastric** and **pyloric**. They secrete two litres of **gastric juice** every day. Mixed with food, this juice forms a milky, semi-fluid paste called **chyme**. The gastric glands produce a secretion with digestive powers.

The pyloric glands make large amounts of mucus, protecting the stomach lining from being digested by its own enzymes. Cells of the gastric glands produce the enzyme, **pepsin**. It is produced as the inactive form, pepsinogen, which only becomes active pepsin when it comes in contact with **hydrochloric acid**.

The hydrochloric acid in the stomach is made by another sort of cell in the gastric glands. The acid kills most bacteria which may be on the food. It also stops salivary amylase from working. So no carbohydrate digestion occurs in the stomach. If there is too much acid in the stomach it can cause the formation of ulcers on the walls.

Pepsin only works if acid is present. It splits protein molecules into smaller units called **peptones**, **proteoses** and **polypeptides**. All of these contain amino acids still linked together.

Babies' stomachs produce the enzyme **rennin**. It converts the protein in milk to **casein**, clotting it as it does so. The milk can therefore remain longer in the gut for proper digestion. As an infant is weaned, pepsin takes over the role of rennin. A muscular area closes the stomach exit until the food is in a semi-liquid state. This region is called the **pyloric sphincter**.

The intestine

From the stomach, chyme enters the **small intestine**. This is seven metres long but gets its name from its small diameter. The first 30cm is called the **duodenum**. The next two and a half metres is the **jejenum**, and the final four metres in the **ileum**. In the small intestine, food breakdown is completed. The bulk of absorption of the digested food is also carried out.

Ducts from the **liver** and the **pancreas** (see Fig. 3.3.6) open into the duodenum. The part played by the liver in digestion is the production of **bile**. It is stored in the **gall bladder**, located just alongside the liver. The gall bladder contracts regularly and empties its bile into the gut when fatty foods are eaten. Bile contains no digestive enzymes. It breaks fat down into small globules because it has chemicals called **bile salts**. These can emulsify fats just like detergents do. About 700 cm^3 of bile is formed each day.

The secretions of the pancreas contain enzymes for digesting all the main types of food substances. The main protein-digesting enzyme is **trypsin**. It changes protein units, proteoses, into smaller peptones and polypeptides. Because trypsin is so powerful, it is vital that it is not activated until it reaches the intestine. If it was activated before, the pancreas itself would be digested. So the pancreas produces **trypsinogen** which is not changed to trypsin until it reaches the duodenum.

For carbohydrate digestion, the pancreatic secretions contain amylase. This changes starch to the smaller maltose molecules. Unlike salivary amylase, it can digest uncooked starch. Fats are already broken up into small globules by the bile salts. The pancreatic fat-splitting enzyme is **lipase**. It breaks fats down into glycerol and fatty acids. They are soon absorbed in this form.

Enzymes secreted by the small intestine complete digestion. They are produced in glands at the end of deep pits called **Crypts of Leiberkuhn**. About three litres of this secretion are produced each day. It is a mixture of enzymes called the **succus entericus**. The enzyme, **enterokinase**, sets trypsin working. Several others complete the breakdown of protein to amino acids.

There are three carbohydrate-digesting enzymes: **lactase** splits lactose (the sugar in milk) to glucose; **maltase** changes maltose to glucose; and **sucrase** splits sucrose to glucose. All these final molecules are small enough for absorption.

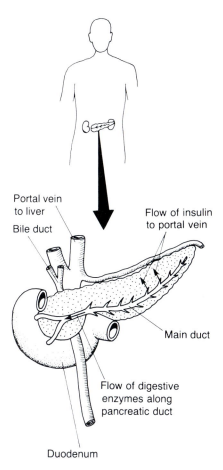

Portal vein to liver

Bile duct

Flow of insulin to portal vein

Main duct

Flow of digestive enzymes along pancreatic duct

Duodenum

Fig. 3.3.6 The position of the pancreas and its ducts

Summary of the stages of digestion

Region	Enzyme	Food substance acted upon	Product
Mouth	Ptyalin (salivary amylase)	Starch	Maltose and dextrin
Stomach	Pepsin	Proteins	Peptones and proteoses
	Rennin	Milk protein (caesin)	Coagulated milk
Duodenum (enzymes secreted by the pancreas)	Lipase	Fats	Fatty acids and glycerol
	Amylase	Starch and dextrin	Maltose
	Trypsinogen (Has to be made active by enterokinase. It then becomes trypsin.)	Proteins and peptones	Amino acids
Ileum (intestinal juice)	Maltase	Maltose	Glucose
	Sucrase	Sucrose	Glucose and fructose
	Lactase	Lactose	Glucose and galactose
	Erepsin (a mixture of enzymes)	Proteoses and peptones	Amino acids
	Enterokinase	Activates trypsinogen	

Absorption The products of digestion are **absorbed** into the blood system of the small intestine. Glucose, amino acids, glycerol and fatty acids, water, minerals and vitamins are all passed through the gut lining. The small intestine has to cope with about 10 litres of fluid every 24 hours. An entirely smooth lining would not be able to complete its absorptive task. The inner surface is deeply furrowed (see Fig. 3.3.7). Millions of small projections called **villi** probe into the gut cavity (see Fig. 3.3.8). There is a maximum density

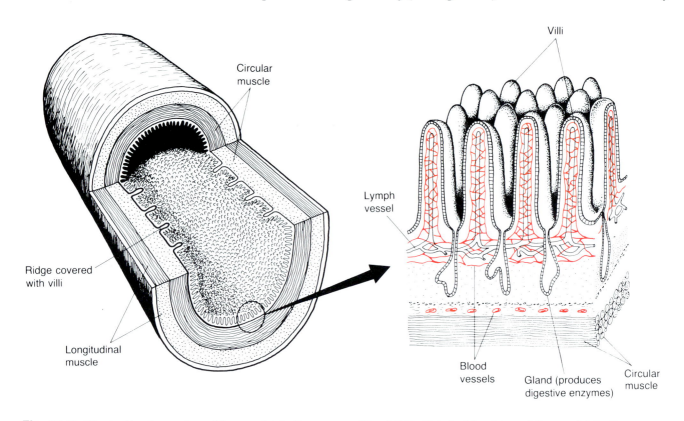

Fig. 3.3.7   The inside of a portion of the small intestine   **Fig. 3.3.8**   Detail of the lining of the small intestine

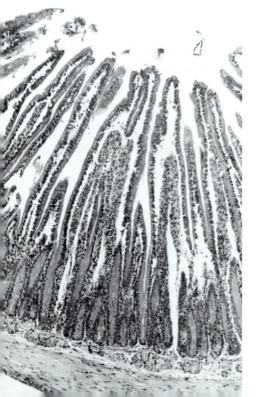

of 40 per mm^2. Each is about 0.5mm long (see Fig. 3.3.9). The villi increase the area for absorption by a hundred times. Also the surface area of the outer column-shaped cells of every villus is fringed with a 'brush border'. There are three thousand projections on some of these cells called **microvilli**. These increase the surface area another 20 times. The total surface area is 550m^2 for the internal surface of the small intestine alone.

Within each villus runs an artery and two veins. These are joined to a network of capillaries (see Fig. 3.3.10). The veins join up with the large portal vein which leads to the liver. In addition, the villus has a central **lacteal** (see Fig. 3.3.11) which is a minute extension of the lymphatic system (see p. 157). During digestion, villi fill with blood and lymph and are swept from side to side. The rippling effect in the massed villi is similar to that produced by the wind in a wheatfield.

At the same time the villi also extend and contract. This movement increases the efficiency of absorption. Absorption of substances in small amounts continues by simple diffusion. The uptake of large amounts uses energy and is called **active transport**. By the time chyme reaches the end of the ileum, 95 per cent of the original 10 litres has been absorbed. The left over half litre contains indigestible matter such as cellulose.

Fig. 3.3.9   Photomicrograph of villi in section

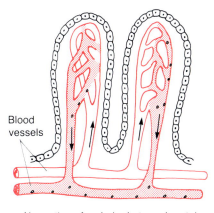

Absorption of carbohydrate and protein

Fig. 3.3.10   Cross-section of two villi

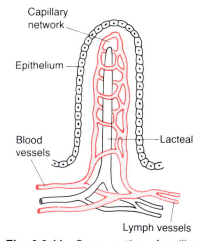

Fig. 3.3.11   Cross-section of a villus showing a lacteal

The final stages

A right-angle bend marks the end of the ileum (see Fig. 3.3.4). Here, at the junction between the ileum and the last part of the alimentary tract, the **large intestine**, is a valve. It is normally kept partially closed to prevent backflow. The large intestine is often called the **bowel**. It is only two metres long but is three or four times wider than the small intestine. Just past and below the valve is a cavity called the **caecum** into which opens the narrow cul-de-sac of the **appendix**. This does not have a function in Man. The caecum continues for a short distance before becoming the **colon**.

The chief tasks of the large intestine are to absorb water plus any useful substances remaining from the food residue. It stores rejectable matter, the faeces, until it can be expelled from the **anus**. All the water absorption in the large intestine is completed by the time the food has travelled half the length of the colon. The lining of the large intestine is capable of absorbing sodium and chloride ions so that a minimal amount of body salt is lost. Millions of bacteria flourish in the colon. They produce vitamin K and vitamin B. These are then absorbed into the blood stream. In the colon mass movements propel the faeces towards the **rectum** and anus. The faeces are mainly water with undigested cellulose, fat and dead bacteria.

The largest gland in the body

Just below your right-hand ribs you have your own incinerator, filter system and storehouse, all rolled into one. This is the liver – the largest single organ in the body (see Fig. 3.3.4). It controls hundreds of chemical reactions needed to make the body work properly. The liver is in the upper part of the abdomen to the right of, and partly above, the stomach.

When a person is at rest a quarter of the body's blood is in the liver. As soon as the person starts physical activity some of his blood is immediately sent to other organs. One of the main tasks of the liver is to deal with chemicals that make up our food. The liver stores some of these. Others, it converts into the sort of chemicals that the body can use straight away.

Some poisonous substances that get into the body can be made harmless in the liver. Many of the medicines we take to cure disease would harm us if they stayed in our bodies after they had done their work. The liver breaks these down into harmless substances. The process is called **detoxification**.

The liver is also responsible for breaking down worn-out blood cells, after they have lived for about 120 days. The iron in them is recycled to make new cells. Vitamins A and B_{12} are stored in the liver.

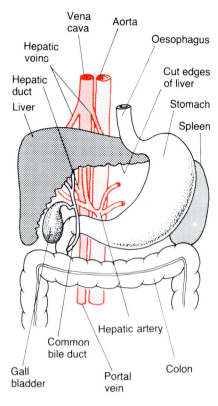

Fig. 3.3.12   The position of the liver and its blood supply

The liver has a double blood supply (see Fig. 3.3.12). The oxygen which its cells need is brought to it through the **hepatic artery**. It also receives blood through the **hepatic portal vein**. This vein carries to the liver all the digested foods that have entered the bloodstream from the small intestine. The blood leaves the liver through the hepatic vein and goes back to the heart.

Foods are taken from the food-rich blood and stored. Molecules of glucose are taken from the blood and joined together in the liver to make **glycogen**. This process is aided by a special chemical, a hormone called **insulin**. If the tissues of the body need glucose to provide energy then some of the glycogen stored in the liver can be turned back into glucose. This process is aided by another hormone called **adrenaline**. The glucose is carried in the blood stream to the muscles or other tissues that need it.

Fat and protein can also be made into glycogen and stored in the liver. Fat can be broken down to provide the body with energy, in the same way as glucose provides energy. When the fat is broken into smaller molecules, these can be built up into glycogen. So, if necessary, the liver can convert fats into carbohydrates to provide the body with energy.

The protein in the diet contains the nitrogen that the body needs. In the intestine proteins are broken down to amino acids. The liver receives many different amino acids in the blood from the portal vein and can carry out different chemical processes with them. If there are too many amino acids, then the liver can break them down and use part of their molecules to build up glycogen. The remainder is waste and is made into the chemical called **urea**. This is passed in the blood to the kidneys and leaves the body in the urine.

The liver cells make bile. It is stored in the gall bladder. Bile contains salts that help in the digestion of fats. It also contains pigments – coloured substances from the breakdown of red blood cells. If the liver is not working properly, too much pigment may be made and jaundice develops. The pigment gives the skin, eyes and urine a strong yellow colour.

Gallstones sometimes form in the gall bladder from cholesterol contained in the bile. These stones can cause great pain and, because they block the bile duct, affect the digestion of fat. Fortunately the gall bladder can be removed without any serious results. The same is not true of the liver. Without it you would soon die.

The liver can be affected by many diseases. **Viral hepatitis** is the most common one. It is highly infectious and destroys the cells of the liver. **Cirrhosis** is another disease of the liver. Excessive drinking of alcohol is one cause of this disease in which the liver cells are destroyed and replaced by fibres of tissue. New cells may grow but the functions of the liver are disturbed in the meantime.

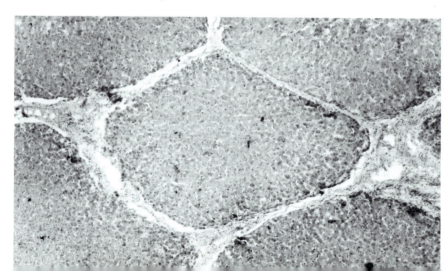

Fig. 3.3.13   A transverse section through a liver lobule

Questions: Domains I and II

1 Gillian and Anne visited the school canteen for lunch. Gillian had vegetable soup, roast beef with potatoes, sprouts and peas followed by a yoghurt. Anne had chips with tomato sauce followed by apple pie and custard.

a Which girl had a balanced meal? Give a reason for your choice. (I)

b Name one food you could add to the main course of the unbalanced meal to improve it. (I)

c Give two reasons why water is needed in the diet. (I)

d Excess fat is stored under the skin. Give two uses of stored fat. (I)

e Excess protein is deaminated. Where does this take place? (I)

f What is the end-product of deamination? (I)

g In what part of the digestive system is the food absorbed? (I)

h Copy out the table and fill in the blanks. (I)

Region	Juice	Enzyme	Food acted upon	Product
mouth	_____	amylase	starch	maltose
stomach	gastric	_____	proteins	
_____	pancreatic lipase		_____	fatty acids and glycerol

2

a Name the structures labelled A–F on the diagram of the alimentary canal. (I)

b For a period of several hours after a meal, samples of food were taken from the parts of the alimentary canal labelled 1–6. The samples were analysed and the results are shown in the table.

Substances	Parts of alimentary canal sampled					
	1	2	3	4	5	6
% starch	8	7	0	0	0	0
% sugars	2	1	8	0	0	0
% proteins	5	1	0	0	0	0
% amino acids	0	2	3	0	0	0
% fats	5	4	2	0	0	0
% water	70	76	76	89	80	70
% cellulose	10	9	9	11	20	30
Enzymes	Carbo-hydrase	Pro-tease	Protease Carbo-hydrase Lipase		None	None
pH	7	3	8	8	8	8

i Humans cannot digest cellulose. How much of the food passing from the mouth is cellulose? (II)

ii These records were taken from a person who was on a sugar-free diet. What evidence is there in the table to show that there must be a carbohydrase in the mouth? (II)

iii Suggest two reasons why there are no amino acids in 1. (I)

iv how much of the sample from 2 is water? (II)

v What has caused the change in the percentage of cellulose between samples 1 and 2? (I)

vi What has caused the change in pH between samples from 1 and 2? (I)

vii What happens to starch between 2 and 3? (I)

viii Which part of the alimentary canal labelled 1–6 contains the largest food molecules? (I)

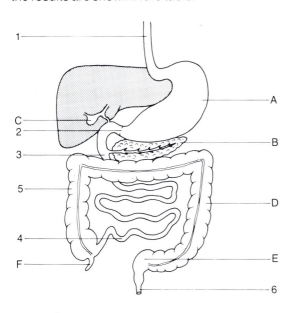

3

a The diagram below shows a section through a villus.

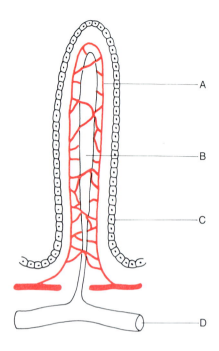

Name the parts labelled A–D. (I)

b In which part of the alimentary canal are villi found? (I)

c Which part of the villus transports glucose away from the alimentary canal? (I)

d Name the two products of fat digestion. (I)

e Which part of the villus transports the products of fat digestion away from the alimentary canal? (I)

f Name two features of villi which make them suitable for the efficient absorption of digested food. (I)

4 The diagram below shows a liver receiving blood from two vessels and passing blood out by one vessel.

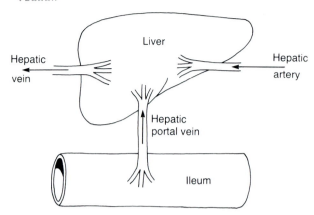

The person has recently eaten a meal containing proteins and carbohydrates.

a Which of the blood vessels will contain the highest percentage of:
(i) oxygen?; (ii) glucose?; (iii) urea? (I)

b Explain why removal of the pancreas might cause death. (I)

5 An experiment was set up to find whether temperature had any effect on the time it would take milk to clot. Eight test tubes, each containing 2cm³ milk and 2cm³ rennin were kept at different temperatures. The times taken for the milk to clot are given in the table.

Temperature °C	0	10	20	30	40	50	60	70
Time taken to clot (minutes)	did not clot	did not clot	10	6	2	4	did not clot	did not clot

a Using the horizontal axis for the temperature, plot a graph of these results. (II)

b Use the graph to find the time taken for the milk to clot at 35°C. (II)

c Explain why the milk did not clot when kept at 10°C and at 60°C. (I)

Experimental skills and investigations: Domain III

1 How does amylase affect starch?

Procedure

1 Label three test tubes 1, 2 and 3. Add 2cm³ of 1% starch suspension to each.
2 Boil 2cm³ of 1% amylase for 1 minute and add it to test tube 1.
3 Add 2cm³ of distilled water to test tube 2.
4 Add 2cm³ of unboiled 1% amylase to test tube 3.
5 Use a teat pipette to withdraw one drop of mixture from each of test tubes 1, 2 and 3.

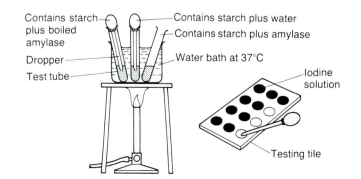

6 Add the drops to iodine solution on a white tile as shown.
7 Record your observations in a table.
8 Place the three test tubes in a water bath at 37°C.
9 Test one drop of each mixture at five minute intervals up to twenty minutes.
10 Record your observations in a table.

QUESTIONS
Using only your observations, state two conclusions of this investigation. Describe how you would detect the presence of the substance produced as a result of the reaction in test tube 3.

2 How does lipase affect milk?

Procedure
1 Add 20cm^3 sterilized milk to a test tube.
2 Add 20cm^3 1% lipase to the test tube plus 1cm^3 phenol red solution and enough 2% sodium carbonate solution to give a reddish tinge.
3 Place the tube in a water bath at 37°C for half an hour.
4 Record any change in the appearance of the contents of the tube.
5 Repeat using boiled lipase.

QUESTIONS
a Given that phenol red changes from red to yellow in acid conditions, explain your observations.
b Suggest why it is important to use sterilized milk rather than fresh milk.

3 How does trypsin affect boiled egg white?

Procedure
1 Add a small quantity (5 or 6 granules) of congo red powder to four drops of water and mix with raw egg white.
2 Suck the mixture into a length of delivery tubing (4mm diameter) and place the length in boiling water for ten minutes.
3 Cool the tubing and push out the solid with a glass rod. (Two eggs can provide enough material for a class of 30 pupils.)
4 Place 2cm of the solid in 2cm^3 of 4% trypsin solution and leave at 37°C for 15 minutes.
5 Record your observations.
6 Repeat using boiled trypsin.

QUESTION
Given that protein digesting enzymes can separate congo red dye from protein, explain fully your observations.

4 How does the lining of your gut work?

Procedure
1 Set up the apparatus as shown in the diagram.

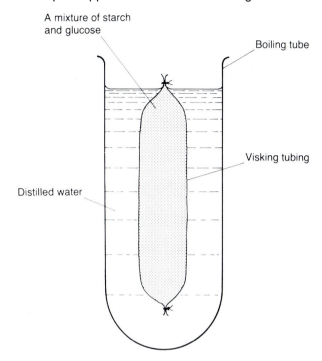

A mixture of starch and glucose

Boiling tube

Visking tubing

Distilled water

2 Test the distilled water with Benedict's solution and with iodine solution as soon as you have set up the apparatus.
3 Repeat the tests at two minute intervals by taking further samples from the boiling tube.
4 Record your observations in the form of a table.

QUESTIONS
a Explain the observations you have made.
b What part of the alimentary canal does the visking tubing represent?
c What does the distilled water represent?
d Explain how the observations illustrate the importance of digestive enzymes.

3.4 The Valentine connection

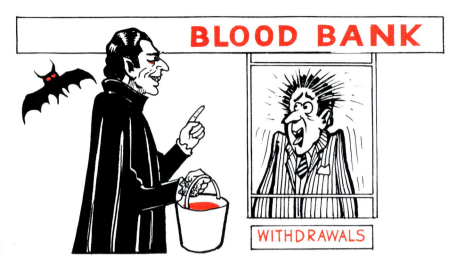

Fig. 3.4.1 Dracula's delight

The heart and circulation

Dracula'a delight – blood

There is a bond that links every man, woman and child on Earth. Every difference of skin colour, religious belief or culture seems small beside it. Some twenty thousand years ago, in a cave, early Man sketched the outlines of a mammoth (see Fig. 3.4.2). He marked the heart in red so making the first known anatomical drawing. Primitive Man had come to understand the link between blood and the heart.

Few words in any language are so full of meaning or have given rise to so many false ideas as those related to blood. 'Blood will tell', 'hot blooded', 'blood is thicker than water', 'it's in the blood' are only a few. Some false beliefs are particularly difficult to correct. It is nonsense to believe that 'Negro blood', 'Jewish blood', or 'Aryan blood' are different from one another. Despite Man's curiosity since pre-history, the discovery of the functions of blood and its circulation had to wait a very long time. It was the Englishman, William Harvey (1578–1657) who made the discovery of the circulation of blood (see Fig. 3.4.3). After twenty years of careful experiment, often using animals, Harvey explained his ideas in a famous lecture. He showed that blood is forced from the right side of the heart through the lungs and from the left side of the heart into the general circulation. Harvey did not really understand the functions of blood. These were not discovered by any single scientist. They were made known, one by one. Various scientists recorded them over the next three hundred years.

Fig. 3.4.2 Cave men drawing mammoths with red hearts

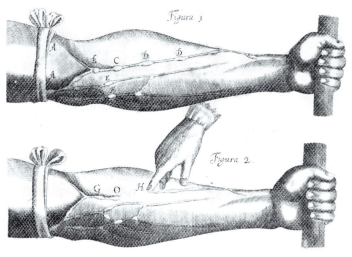

Fig. 3.4.3 William Harvey and his discovery of the circulation of the blood

If blood is left to stand in a container, a thick red mass of cells settles to the bottom. A clear liquid, the **plasma**, appears on top of the mass (see Fig. 3.4.4). Plasma is ninety per cent water, the other ten per cent being mainly mineral salts. In this liquid most of the substances that need to be carried around the body are dissolved and the blood cells (**corpuscles**) float.

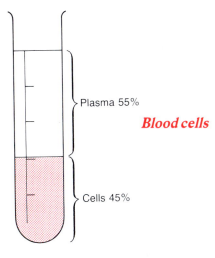

Plasma 55%

Cells 45%

Fig. 3.4.4 Blood separates if left to stand in a boiling tube

Blood cells

Each second of your life two and half million **red blood cells**, 120 000 **white blood cells** and five million **platelets** are formed (see Fig. 3.4.5). They are needed to replace those that have worn out. In an embryo, blood is made almost entirely in the liver and spleen (see Fig. 3.4.6) but in an adult blood production is confined to other parts of the body. The red marrow of some bones makes red cells, platelets and some types of white blood cells. The spleen and lymph glands supply other types of white cells called lymphocytes.

Before joining the blood stream, each red cell and white cell (polymorph) goes through a complex growth period within the bone marrow. Sometimes something goes wrong in the blood. A sample is taken of the blood-producing marrow from the breast bone. This allows biologists to estimate what has gone wrong in the development of the cells.

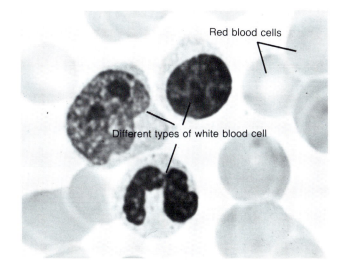

Red blood cells

Different types of white blood cell

Fig. 3.4.5 Photomicrograph of blood cells

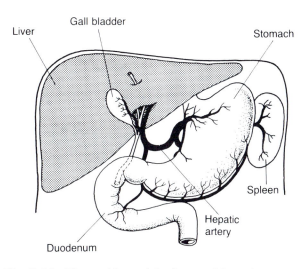

Gall bladder

Liver

Stomach

Spleen

Hepatic artery

Duodenum

Fig. 3.4.6 The positions of the liver and the spleen

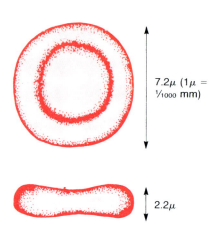

7.2μ (1μ = 1/1000 mm)

2.2μ

Fig. 3.4.7   Red blood cells

Red cells
Each is a cell saturated with **haemoglobin**, a reddish substance rich in iron. The iron allows the cell to hold a small quantity of oxygen when it passes through the lungs. It carries this to the tissues. In its three months of life, the red cell will travel about 700 miles. It is then taken out of the circulation by the spleen. organ breaks up the battered, worn-out cells. The iron is recycled to be used again in the bone marrow.

The rest of the material is sent to the kidneys and other organs to be eliminated. The waste products of the red cells are changed by the liver into the greenish-coloured **bilirubin**. It is secreted in the bile.

Red cells do not have nuclei. They are shaped like flattened discs with a depression on either side (see Fig. 3.4.7). In one mm^3 of blood there are about five million red cells. This means that Man's body, with its five or six litres of blood, contains over 25 billion red cells. Laid side by side they would reach four and a half times around the world! (See Fig. 3.4.8.)

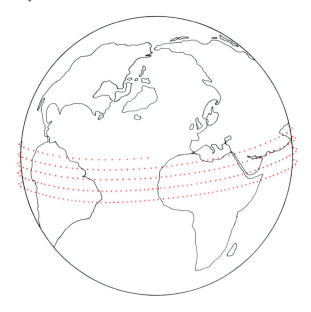

Fig. 3.4.8   Man's red blood cells would reach four and a half times around the world

White cells
These have nuclei and are larger than red cells. About 70 per cent have irregularly shaped nuclei and are called **polymorphs**. The other 30 per cent, with round nuclei, are called **lymphocytes**. One mm^3 of blood contains about 6000 white blood cells.

Platelets
These are small flat oval bits of larger cells formed in the bone marrow. There are about 300 000 per mm^3 of blood. They help to clot and live for about nine days.

The great carrier   Without blood, the body would not be able to supply cells with their vital needs or remove their harmful waste products (see Fig. 3.4.9). In order to release energy, cells need a fuel and oxygen in which to 'burn' the fuel. After the release of energy from this fuel, useless material in the form of waste remains. Useful materials supplied to the cells include fuel in the form of products of digestion. Oxygen is supplied by the red cells.

Products of digestion include:

a   glucose (carbohydrate);
b   amino acids (protein);
c   glycerol and fatty acids (fats).

These are all carried in solution in the plasma together with vitamins and minerals. Fats, glycerol and fatty acids are also carried by lymph (see p. 155). Oxygen adds on to haemoglobin in the lungs where there is a large proportion of oxygen in the air breathed in (about 20 per cent). The haemoglobin then becomes **oxyhaemoglobin** which is bright pinkish red in colour.

HAEMOGLOBIN + OXYGEN = OXYHAEMOGLOBIN

(in the lungs)

The oxyhaemoglobin returns to the heart and is pumped to all the tissues that are in need of oxygen. In these tissues, the proportion of oxygen in the dissolved air around them is low. Because of this, the oxyhaemoglobin breaks down and gives up its oxygen to the cells, leaving haemoglobin to be recycled and used again (for about three months).

OXYHAEMOGLOBIN = OXYGEN + HAEMOGLOBIN

(in body cells)

When energy is released in cells during respiration, the main waste product is **carbon dioxide**. If it stayed in or near the cells it would react with water to form **carbonic acid**. This would harm the cells. It has to be removed and taken back to the lungs where it can be eliminated. Some carbon dioxide dissolves in the plasma and is carried in this way. Most carbon dioxide is transported as **sodium hydrogen carbonate.**

Carbon dioxide produced in the cells reacts with water to form carbonic acid. A complex series of reactions then takes place. Eventually sodium, from sodium chloride (common salt) in your diet, adds on to hydrogen carbonate from carbonic acid. This reaction results in sodium hydrogen carbonate being formed. It dissolves in the plasma and is carried to the lungs after passing through the heart. At the lungs it breaks down to give carbon dioxide which is breathed out.

CARBON DIOXIDE (from cells) + WATER = CARBONIC ACID

CARBONIC ACID forms HYDROGEN CARBONATE

SODIUM (from sodium chloride in your diet) +
    HYDROGEN CARBONATE =
                    SODIUM HYDROGEN CARBONATE

SODIUM HYDROGEN CARBONATE breaks down in the lungs
    to give CARBON DIOXIDE

Another waste product which is formed in the cells of the liver is **urea**. It forms from amino acids which are not wanted by the body (see section on excretion p. 173). The urea leaves the liver, dissolved in plasma, in a vein (see p. 132) and joins the general circulation. Eventually it passes into the kidneys and is filtered out of the blood when urine is formed.

Besides the products of digestion, other useful materials carried by the blood include **hormones** (see p. 188). Briefly, these are chemical messengers which tell parts of the body what to do. They are made in many glands of the body called **endocrine glands**. These glands do not have tubes (ducts) leading from them. The hormones must be carried away by the blood system.

Fig. 3.4.9 'Mr Blood the porter'

Keeping an even temperature

Blood is essential for keeping the body within the narrow limits of temperature necessary for life. Blood transports water to the sweat glands thus helping to cool the body. It carries heat from the liver and muscles to the skin surface where it is radiated to the air. Also, dilation and constriction of blood vessels help to keep the body temperature constant. These mechanisms are regulated by a thermostat in the brain. When the body's temperature drops below, or rises much above normal, it is a signal indicating illness.

On the defensive

White blood cells combat one of our most dangerous and unconquered enemies – disease-causing micro-organisms (see Fig. 3.4.10). Red cells out-number white cells by about 600 to one, but have a completely different function. As soon as hostile micro-organisms enter the body we begin to fight back. We do this in a number of ways, each usually involving the action of one or more types of white blood cell. Some types pass through the walls of the smallest blood vessels (see Fig. 3.4.11). They move like certain animals made of one cell, for example, *Amoeba*. Once they make contact with the alien invaders they take them into their bodies and digest them. Often the invaders produce poisons.

Other types of white blood cells produce anti-toxins which cancel the action of the poisons. Yet other white cells are able to produce special proteins called **antibodies**. These immobilize the enemy while they are being digested. Sometimes the attack is too violent even for all the types of white cells. When this happens we become ill, often needing help from the doctor's antibiotics (see p. 47).

The ability of the blood to from a clot so that it stops flowing is an important part of our defence system. The clot will prevent blood loss as well as helping to prevent entry of foreign micro-organisms. Normally when blood is exposed to air it hardens into a gelatinous mass. When this mass is examined under a microscope, it appears as a lacey network of fine rods holding the blood cells together. It looks like a fisherman's net holding a wriggling catch of sardines and squeezing them together to make an almost solid block of fish. The 'fish-net' is called **fibrin** and the process of its formation is very complex. It is simplified and summarized in Fig. 3.4.12.

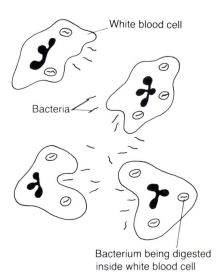

Fig. 3.4.10   White blood cells ingesting bacteria

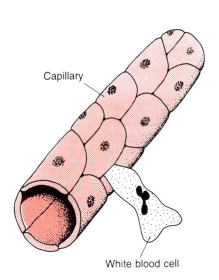

Fig. 3.4.11   A white blood cell escaping through a capillary

Fig. 3.4.12   Formation of a blood clot

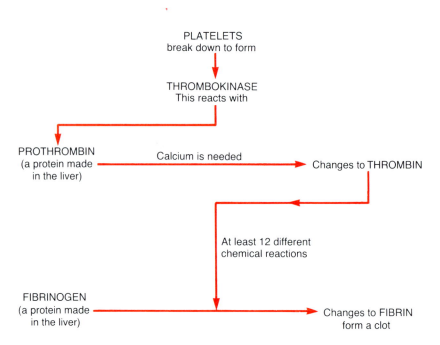

Individual treatment – blood groups

Over many centuries, we have developed ways to protect ourselves against attack. Weapons, warm clothes, and fire were effective against many menaces from outside. The body and the blood were necessary in the fight against the silent, hidden threat from foreign bodies which penetrated the skin's defence. After Harvey's discovery of the circulation of the blood, various attempts were made at transfusion (see Fig. 3.4.13).

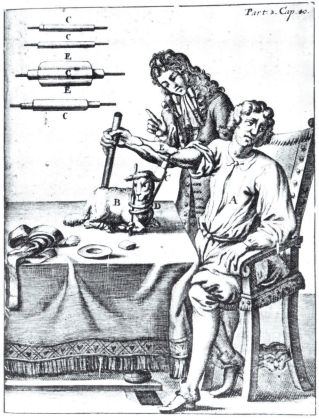

Fig. 3.4.13 A man receiving a blood transfusion from an animal in the seventeenth century

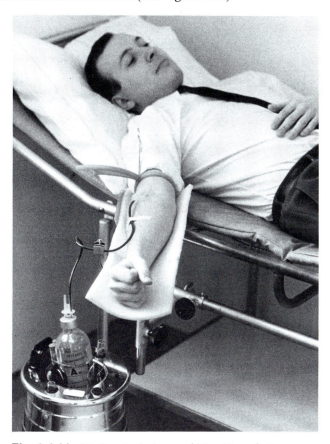

Fig. 3.4.14 Modern techniques of blood transfusion

Animals were used at first and almost all those who were given animal blood died. Only in 1901 was the discovery made which changed desperate hit-or-miss techniques into one of the surest aids known to medicine (see Fig. 3.4.14). Dr Karl Landsteiner of Vienna used his observations made on animal blood to lay the foundations of blood grouping in humans. He showed that mixing different types of blood caused the cells to clump together (**agglutination**). He showed that this often happened with human blood. Then, working with a large series of samples, he recorded the basic groups of blood. Within a few years he arrived at the A, B, AB, and O classification which has formed the basis of blood-typing ever since.

Agglutination is caused by a reaction between proteins called **antigens** made on red blood cells and proteins called antibodies in the plasma. If you belong to group A, you have antigen A in your red cells; a group B person has B antigen; AB has both A and B antigens; while group O has neither of them.

The antigens can be identified by their reaction with two antibodies (**anti-A** and **anti-B**). Anti-A will clump together red cells containing A (Groups A and AB), while anti-B will clump together red cells containing antigen B (Groups B and AB). Neither of them will react with Group O.

Antibodies develop shortly after birth. When a blood transfusion is needed, the red cells of the blood selected must belong to a group which will not be affected by any antibody in the patient's plasma (see Fig. 3.4.15).

Those who can safely receive blood of each type		Those who cannot
O A B AB	Type O	
A AB	Type A	O B
B AB	Type B	O A
AB	Type AB	O A B

Blood group of recipient	Donor's blood group				
	Group O	Group A	Group B	Group AB	✓ Safe transfusion (compatible bloods)
Group O	✓	●	●	●	
Group A	✓	✓	●	●	
Group B	✓	●	✓	●	● Dangerous transfusion (incompatible bloods)
Group AB	✓	✓	✓	✓	

Fig. 3.4.15 Charts to show possible blood transfusions (ABO grouping)

After the discovery of the ABO blood grouping system a major problem remained. This was to prevent the blood from clotting before it could be given to the person who needed it. In the first decade of this century, transfusions were given directly from arm to arm (see Fig. 3.4.16). A rubber tube was used but tiny clots were likely to form in the tube. Also this method was a way of transmitting certain infectious diseases.

Fig. 3.4.16 An early attempt at arm to arm transfusion

In 1913, a Belgian doctor, Albert Hustin, discovered that adding a small amount of sodium citrate and citric acid would prevent blood clotting. The sodium citrate reacted with calcium in the blood (see p. 140). so that calcium was not available any more for the blood clotting process. Today, **heparin** is added to blood in blood banks to prevent clotting. Indeed, it is produced by many blood-feeding animals to maintain a constant flow of blood from a wound.

In blood banks glucose is added to 'feed' the red cells while in storage. Most blood donated to blood banks is refrigerated at 4 degrees C and can remain useful for up to 18 days. Plasma, however, can be recovered and kept almost indefinitely (see Fig. 3.4.17).

The treacherous Rh factor   Since Landsteiner's discovery of the four main blood groups, many other sub-groups have been found. Most are of little importance in daily practice. There is an exception: the Rh factor discovered by Landsteiner and his colleagues shortly before the Second World War. While studying the blood of the Rhesus monkey (see Fig. 3.4.18), they found a substance previously unknown and called it the **Rhesus factor**. When testing humans, 85 per cent were found to be **Rh positive** (which meant that their blood cells clumped together in the presence of this factor) and the remaining 15 per cent were **Rh negative**. This led directly to solving the mystery of why some mothers bore one or two healthy children and then had babies with symptoms of agglutination and anaemia (see p. 145). The babies often died because the mother was actively producing antibodies that were destroying them.

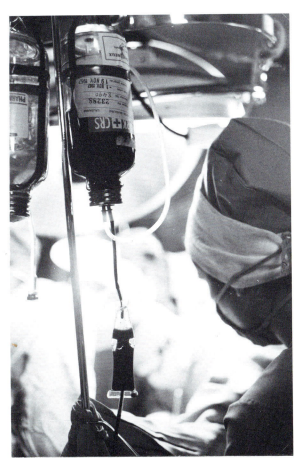

Fig. 3.4.17   Photograph of modern transfusion bottle being used

Fig. 3.4.18   A Rhesus monkey

Although mother and unborn baby have separate circulatory systems, some leakage does occur. When Rh positive cells from the embryo leak across the placenta into the mother's blood (Rh negative) she produces antibodies against them. If these antibodies are diffused back into the embryo, damage and sometimes death are inevitable. The first baby is rarely affected because the mother's body needs extra time to build up a dangerous quantity of antibodies. Once this process has begun in the mother, the production of these antibodies continues so that a second Rh positive child may be doomed.

Modern methods of treatment vary. Severely affected newborns can have their lives saved by a complete blood transfusion, draining off the Rh positive blood. As it grows older the child will again make blood of its own type but by that time the danger from the mother's antibodies will be past.

Another method is intra-uterine transfusion. A tube is inserted into the baby and blood is transfused while the baby is still in the womb. Best of all, there is a way to prevent the danger itself. Within 60 hours after Rh negative women give birth to their first baby, they are given the anti-Rh antibody to prevent them from making antibodies. In 1965 the first Rh positive baby was born after the mother had been given this treatment. It was free of Rhesus disease. Many countries now have large-scale programmes to provide enough anti-Rh antibody for all the women who need it.

How your blood group places you

In different parts of Britain the percentages of the four blood groups, A, B, AB, and O vary considerably. When countries or continents are compared, the differences can be very large. These differences in blood group percentages last for many generations. For example, gypsies in Europe have blood groups like those of the population of North-West India, where they are thought to have come from. It is thought that the earliest inhabitants of the British Isles were largely in group O. Their

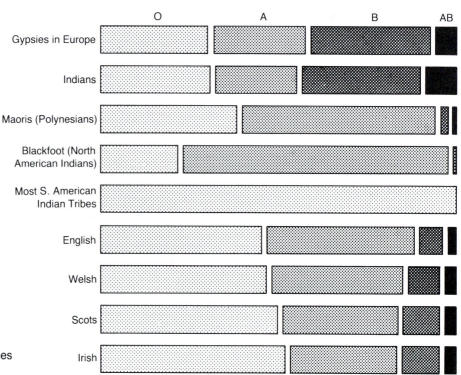

Fig. 3.4.19 Approximate percentages of blood groups in some races

descendants were pushed north and west by invading Anglo-Saxons who had a greater percentage of group A. The approximate percentages of blood groups in some races are given in Fig. 3.4.19.

When something goes wrong

Problems often occur if the quality or the quantity of blood cells are altered. (Parasitic diseases of the blood are considered on pp. 39 – 40). There are three major groups of blood disorders:

1 **Anaemia**, resulting from an insufficient number of red cells or a low content of haemoglobin.

2 **Leukaemia**, when white cells multiply abnormally.

3 **Haemophilia**, due to a breakdown in the clotting process.

Anaemia: lack of iron

The first signs of anaemia are usually a feeling of tiredness and paleness of the skin. Its main cause is lack of iron, which robs the red cells of their colour. Even if the red marrow goes on making red cells they do not contain enough haemoglobin to replace the iron the body uses. Giving a person large doses of iron will usually cure this type of anaemia. Meat, eggs and some fruits are rich in iron. Consequently, anaemia is now rare in the more wealthy countries of the world.

Sometimes the need for iron is greater than the normal supply. When a woman is pregnant, for example, the baby draws heavily on her reserve of iron. Often after repeated pregnancies, the stock runs out and extra iron is needed in the diet. Pernicious anaemia, which used to be fatal, is a result of lack of vitamin B_{12}. It results from the bone marrow not being able to make normal red cells. Instead it wears itself out producing a small number of abnormally large red cells that are bloated with haemoglobin. In most cases it can be cured by injections of 1000th of a milligram of vitamin B_{12}.

Sometimes the inability of red cells to carry enough oxygen can be inherited. One of the best known diseases of this kind is sickle-cell anaemia. It shows itself in the form of abnormally shaped red blood cells and a proportion of haemoglobin which cannot carry enough oxygen. However, people who suffer from this problem tend to be immune from malaria and so those who live in areas where malaria occurs may be at a certain advantage. If they survive the effects of limited oxygen supply they will also be able to survive in parts of the world where malaria occurs. See also the section on genetics (p. 212).

Leukaemia: white cells out of control

Leukaemia is caused by an abnormal increase of white blood cells. The bone marrow begins producing masses of these cells in an uncontrolled way. An enormous increase in the production of polymorphs occurs and they enter the bloodstream at every stage in their development. This means that 1 mm^3 of blood contains hundreds of thousands of polymorphs of all kinds instead of a few thousand fully developed ones. The marrow often produces white cells which, because of their monstrous size, resemble cancerous cells.

This leaves little room in the marrow for the production of normal red cells and platelets. To overcome this problem, grafts of normal red bone marrow are given to the patient but, like other grafts, they may be rejected. To lower this risk, donors are carefully chosen to match the patient's tissues as closely as possible and all of the patient's defences are first suppressed.

Fig. 3.4.20 A royal complaint!

Haemophilia: a royal complaint

Two sex chromosomes determine whether a child is a girl or a boy. One is provided by the mother, the other by the father. If there are two X chromosomes a girl is born. If the mother's chromosome X joins with the father's Y chromosome, it is a boy. Haemophilia is a hereditary disease (see pp. 211 – 213) transmitted by an abnormal X chromosome. It prevents the blood from clotting properly when it comes in contact with air. A girl will not be directly affected by haemophilia since she always has two X chromosomes and one of these will almost invariably be normal.

The normal X chromosome over-rules the effects of the abnormal one but she is still able to transmit the problem to some of her children (see p. 213). When an abnormal X chromosome meets a Y chromosome, the result is a boy with haemophilia. His sons will not suffer from it because they will inherit a normal Y chromosome but his daughters may transmit the disease because of the defective X chromosome they have received from him.

Haemophilia helped shape the history of Europe because it involved members of British royalty. Queen Victoria passed it on to the reigning families of Spain and Russia in the beginning of the twentieth century. Because of her son's haemophilia, the Tsarina of Russia allowed Rasputin to influence his country's foreign and internal policies. The Tsarina thought that Rasputin had magical powers that could cure her sons problem.

Today, when a haemophiliac needs an operation of any sort, he is first given plasma extract containing a concentration of anti-haemophilia substances. Despite this treatment, haemophiliac children must be protected throughout their lives from blows, injury or violent effort. They are usually advised to find an indoor profession, for example, an office job.

The circulatory system: pulsating network of life

During the average person's lifetime the work done by the heart is equal to that needed to raise their whole body weight to a height of over a thousand kilometres! All this work pumps blood in an endless circuit around the body, minute after minute, for year after year.

Besides a pump, the circulatory system consists of a double network of branching tubes through which the blood passes. One of these networks supplies the body with blood which is rich in oxygen. It returns blood, loaded with carbon dioxide, to the heart. This is the **systemic circulation** (see Fig. 3.4.21). The other takes blood in a circuit between the heart and lungs. In the lungs the blood takes up oxygen from the air

Fig. 3.4.21 Plan of pulmonary and systemic circulations

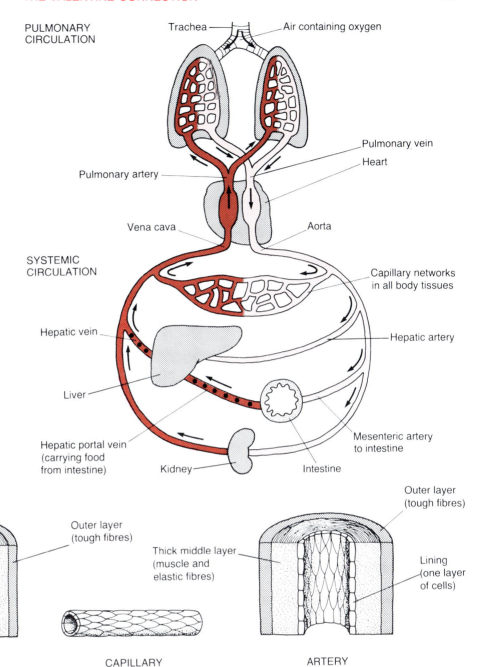

Fig. 3.4.22 Sections through a vein, capillary and artery

and releases carbon dioxide into the air. This system is the **pulmonary circulation** (see Fig. 3.4.21). Blood is carried from the heart in vessels called **arteries** and returned to the heart in **veins** (see Fig. 3.4.22). The output side (**arterial**, see Fig. 3.4.23) is connected to the input side (**venous**, see Fig. 3.4.24) at very fine endings of the arterial and venous systems by tubes called **capillaries** (see Fig. 3.4.22). These very small tubes have walls only one cell thick which enable small molecules to pass freely across them.

Blood returns to the heart in veins. This happens partly by gravity (in the upper part of the body) and partly by pressure of muscles as they contract and relax near veins. Special valves in the veins keep blood flowing in one direction only – back to the heart. So in this respect the body muscles, helped by valves in the veins, act as a return pumping system. In every region of the body below the heart, the pump must work against gravity.

Fig. 3.4.23   The arterial system

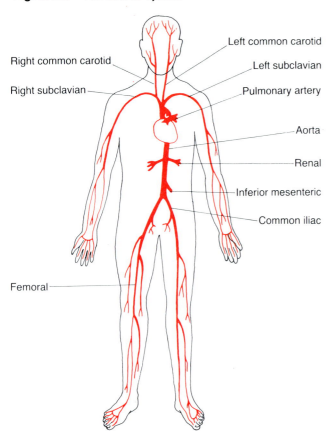

Right common carotid
Right subclavian
Left common carotid
Left subclavian
Pulmonary artery
Aorta
Renal
Inferior mesenteric
Common iliac
Femoral

Fig. 3.4.24   The venous system

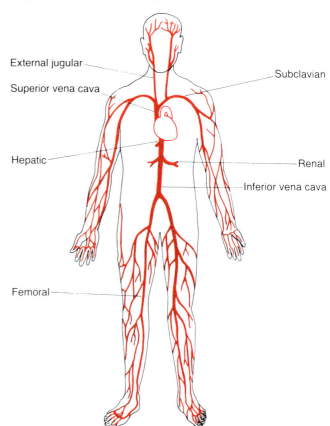

External jugular
Superior vena cava
Subclavian
Hepatic
Renal
Inferior vena cava
Femoral

The amazing machine

'I want you to design a pump,' the Chairman said to the Production Engineer.

'That's what I'm here for', the Production Engineer replied confidently.

'Now listen to me,' the Chairman went on. 'The pump must keep eight pints of fluid on continuous circulation at a temperature of 37 degrees C. It must work against constantly changing resistance and must adapt itself in an instant, sometimes pumping eight pints per minute, sometimes fifty.'

'What do you mean by in an instant?'

– 'Within a tenth of a second.'

The Production Engineer frowned and gave a whistle through his pursed lips.

– 'The pump must weigh no more than 350 grams. It must work at the height of Mount Everest, or in the Sahara desert, or at the North Pole. It must work by day and night and at 130 strokes per minute – then in a couple of minutes, drop to seventy. It will be controlled by a self-regulating, electrical mechanism.'

– 'You mean automation?' the Production Engineer asked. 'We are quite used to that.'

– 'The pump must also respond to human control – which at any time may counteract the automation. The controller of the pump may be asleep, or drunk – it makes no difference.'

– 'It's going to be hard to design a pump like that.'

– 'Another thing. The pump has four chambers with four valves, and has to drive the fluid in opposite directions at the same time.'

– 'What about running repairs?'

– 'Servicing will have to be done without the pump losing a single stroke.

Fig. 3.4.25 'The amazing machine'

The pump must never stop. Oh, just one other minor point. The pump must work for anything up to 100 years.'

At this point, the Production Engineer let his pencil fall. 'It can't be done sir! A pump like that simply could not be designed.'

Then the Production Engineer saw a smile of triumph on the Chairman's face.

– 'Oh I see. It is an imaginary pump; it's your dream – for the future.'

– 'No, it is a real pump. You have one under your shirt. The finest pump that was ever invented, it grew on its own. Mine's been going for half a century – with automatic servicing. There has never been anything like the human heart – it's nothing but a pump with tubes attached. Four chambers, two circulations, all kept up with perfect automation. And it even works when we are not thinking. It can respond to any requirement – heat and cold – bad temper, beating the mile record. It is a lesson in design, in servicing – everything.'

– 'But the real point is' – here the Chairman paused – 'The heart does go wrong. The material wears out before it should. Those tubes become rough inside and are silted up. The valves get glued up. The automation goes wrong and we simply don't know enough about how the pump works, to put it right.'

The Production Engineer became more hopeful – 'Surely sir, that's a subject for study?'

– 'It surely is. To get to the bottom of that problem we shall need to spend ten million pounds a year for many years, on research.'

The above conversation shows just how remarkable the heart is and that we still have much to learn about it. However, there are certain facts that we do know and a knowledge of these is essential if we are to understand the heart's action.

The parts of the heart
(Fig. 3.4.26)

Each side, right and left, has a collecting chamber, the **atrium**, to receive blood, and a pump, the **ventricle**, to pump it out. Reverse flow is prevented by valves between ventricles and atria and between ventricles and the arteries which take blood from the heart. Thin strands of muscle hold the valve flaps to the walls of the ventricles and prevent the flaps from being pushed back into the atria under pressure. The valves guarding the **aorta** and pulmonary arteries have three flaps each. The left ventricle has a much thicker wall than the right and is circular in cross-section, the septum bulging into the right chamber. The reason for the difference is that, though they both pump the same amount of blood, it takes much less effort to pump blood to the near-by lungs than to the whole of the rest of the body.

RIGHT SIDE   LEFT SIDE

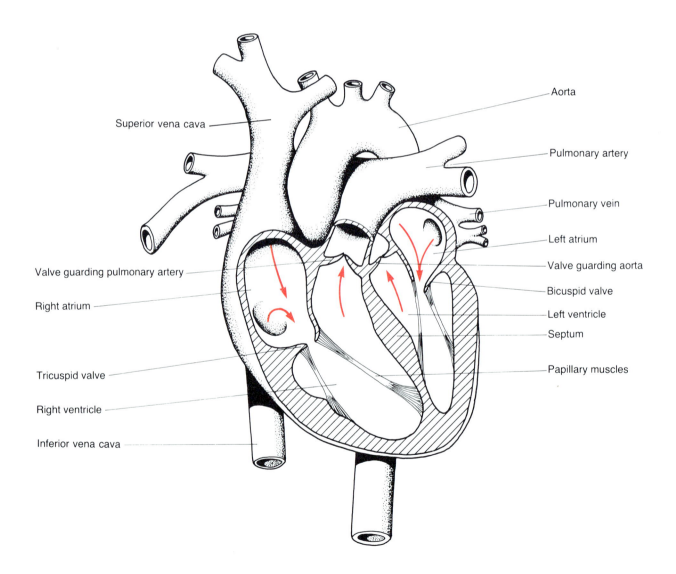

Fig. 3.4.26   Cross-section of the heart

It is possible to summarize the path taken by blood as it passes twice through the heart (double circulation).

1 VENAE CAVAE
   (carry blood with
   very little oxygen
   from body tissues).

   ↓

2 RIGHT ATRIUM

   ↓

3 TRICUSPID VALVE

   ↓

4 RIGHT VENTRICLE

   ↓

5 PULMONARY ARTERIES
   (carry blood to lungs to
   collect oxygen).

   ↓

6 PULMONARY VEINS
   (carry oxygenated blood
   back to heart).

   ↓

7 LEFT ATRIUM

   ↓

8 BICUSPID VALVE

   ↓

9 LEFT VENTRICLE

   ↓

10 AORTA
   (carries blood with
   a lot of oxygen
   around the body).

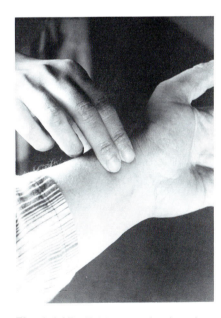

Fig. 3.4.27 Taking a patient's pulse

The steady beat When a doctor calls on a patient, his first step is often to take the patient's pulse. Holding the patient's wrist on the side near the thumb, he obtains his first information about the heart (see Fig. 3.4.27). A man's normal pulse rate is about 60 to 70 beats per minute. For a woman it is about 78 to 82 beats per minute. Children have the highest rate of all: about 130 beats per minute at birth. Blood or arterial pressure is the pressure at which blood circulates through the arteries (see Fig. 3.4.28). A doctor or nurse takes a patient's blood pressure by finding out how much pressure is required to stop the flow of blood in the arm (see Fig. 3.4.29). A strap is tied around the arm. A blood pressure reading gives two pieces of information. The maximum pressure corresponds to the amount the heart is able to push out against resistance. The minimum pressure is the amount the heart pumps normally when resting.

'Calm down, remember your blood pressure.' 'He makes my blood boil.' These everyday expressions show how much all of us are aware of the effect that emotion has on the rate and pressure of blood.

Fig. 3.4.28   An early attempt at taking blood pressure in an animal

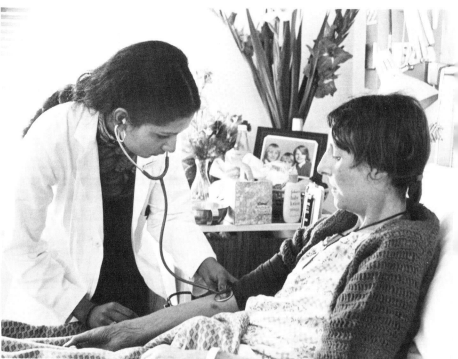

Fig. 3.4.29   A nurse taking a patient's pulse

You are as old as your arteries

Blood pressure varies according to the amount pumped out by the heart and the resistance it meets in the blood vessels, particularly in the arteries. Young people have supple, elastic arteries and blood moves easily through them, helped by the recoil of the arteries themselves (see Fig. 3.4.30). As a person grows older, this elasticity is lost and more resistance is offered to the flow of blood thus forcing the heart to work harder. Arteries become like an old rubber hose, brittle and fragile.

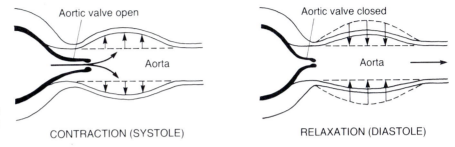

CONTRACTION (SYSTOLE)   RELAXATION (DIASTOLE)

Fig. 3.4.30 Changes in shape of the elastic arterial walls during contraction and relaxation

When arteries become hard, or sclerotic, they are unable to expand. Pressure begins to build up in them. High blood pressure often reveals itself by causing headaches, ringing in the ears, dizziness, cramps and tingling sensations in the fingers. The person has trouble seeing and is often bothered by flashing lights or spots before the eyes. Nose bleeds are other symptoms. In the most serious cases, some arteries in the brain may burst. This is called **apoplexy**.

Nervous control of the heart

In the wall of the right atrium, near the entrance of one of the venae cavae, is a bundle of nerve tissue called **a node** (the **sinu-atrial** or **SA node**). Here the nervous system begins to control the rate of heart beat. For this reason the SA node is called the **pacemaker**. When the nervous impulse begins, it starts a wave of contraction in the heart muscle. It spreads out over the right and left atria. There is another pacemaker (the **atrio-ventricular** or **AV node**) in the wall between the two sides of the heart. From this, a bundle of nerves called the **bundle of His**, extends out to reach both ventricles. There is a network of nerves which spreads out to start contractions from the bottom of the ventricles. These nerve fibres are the **Purkinje fibres**.

The heart beat sounds

The stage during which the ventricles are contracting is called **systole**. The period during which the heart is relaxed and the ventricles are filling with blood is called **diastole**. By using a stethoscope the sounds made by the heart can be made louder. There are two distant sounds. The first, a 'lub', is the loudest and longest. The second is quieter and shorter – a 'dub' sound. The sounds are rhythmically repeated in a 'lub-dub', 'lub-dub' pattern. The first sound (the 'lub') is made by the sudden closing of the valves between the atria and ventricles and the opening of the valves guarding the aorta and pulmonary artery. The second sound (the 'dub') is made by the closing of the valves guarding the aorta and pulmonary artery and the opening of the valves between the atria and ventricles.

The whole heart beat takes less than one second. The nerve impulses can be recorded as an **electrocardiogram** which is really a graph showing the electrical changes in the heart against time. It records the small electrical changes that occur in the muscles of the heart when they contract. Figure 3.4.31 shows an electrocardiogram next to a diagram of the pacemaker system of the heart. The recording of the electrical change represents nerve impulses passing through the heart muscle.

Fig. 3.4.31 An explanation of the electrocardiogram

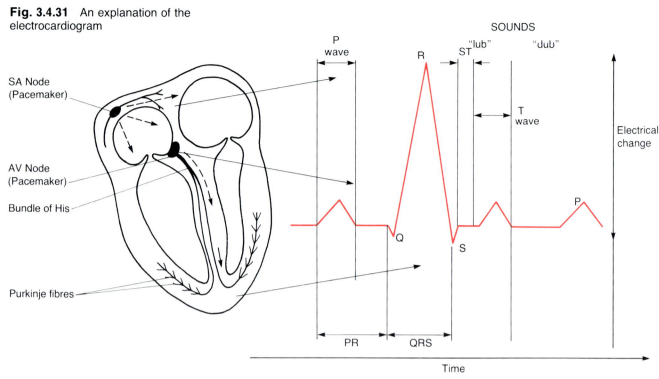

Explanation of Fig. 3.4.31

The *P* wave shows the passing of the nerve impulses from the pacemaker in the atrium through both atria.

The *PR* part shows the contraction of the atria and the passing of the nerve impulse to the ventricles.

The *QRS* part shows the passing of the nerve impulse through the ventricles.

The *ST* section shows the contraction of the ventricles.

The *T* to *P* wave shows the final relaxing of the ventricles.

Heart disease   The most common reason why the heart stops working is that it runs out of oxygen. Oxygen is needed to help release energy for the contraction of heart muscle. The heart gets its oxygenated blood through the **coronary artery** which is the first branch of the aorta. If the coronary artery becomes blocked, the oxygen supply is cut off and the heart soon stops beating. This is called **a heart attack**. It is the most common cause of death in Britian.

Some causes of heart attack

1   Blockage of the coronary artery with
a   **Cholesterol**. This is a fatty substance which lines the inside of arteries if there is too much in the diet.
b   **Blood clots**. Due to abnormal conditions, blood clots can block the coronarty artery.
2   **Abnormally high blood pressure**. This can be caused by worry, being overweight, smoking and drinking alcohol. The harder the heart works, the more oxygen it needs and the greater the risk that this oxygen supply may fail.

Most people who have heart attacks have some combination of the above factors. Many years of not looking after your fitness can lead to a heart attack. The only way to reduce the chances of such a heart attack is to avoid as many of the causes as possible.

The medium between blood and cells

Most body cells are bathed in a fluid which acts as a medium between them and the blood. It is necessary because whole blood with dissolved materials never leaves undamaged blood vessels. Blood is under pressure in the smallest branches of arteries so plasma is forced through the walls of the capillaries. Vital materials are therefore able to reach cells. Waste materials and hormones produced by cells dissolve in the bathing plasma which then carries them into capillaries. When outside the blood vessels, the plasma surrounding cells is called **tissue fluid**. See Fig. 3.4.32.

As a result of continuous seepage through the capillaries there is a danger of tissue fluid collecting and remaining around cells. In order to prevent this, there is a series of thin-walled tubes draining the fluid towards the blood system. The tubes form a branching network leading from most parts of the body and make up the **lymphatic system** (see Fig. 3.4.33). Its fine branches lead to larger ones which drain into the main one called the **thoracic duct**. Eventually, the thoracic duct joins a major vein leading from the left arm. In this way, tissue fluid is returned to the blood system.

Fig. 3.4.32 The formation of lymph

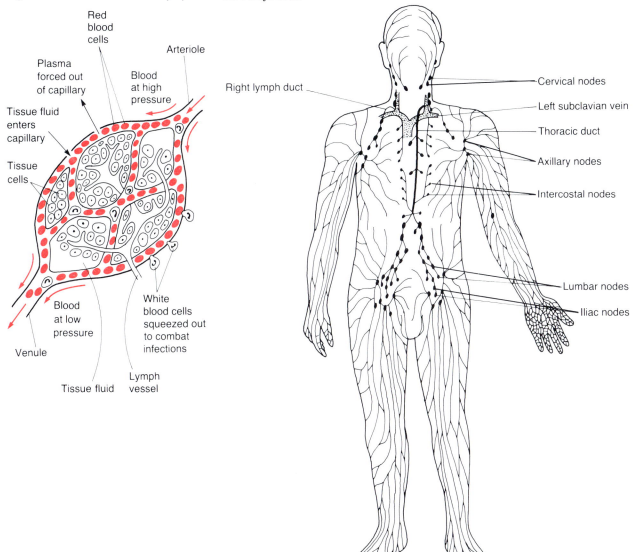

Fig. 3.4.33 The lymphatic system

At certain points along the **lymph vessels** are swellings called **lymph nodes** or glands (see Figs 3.4.34 and 3.4.35). Within each lymph node are formed lymph cells called **lymphocytes**. They divide repeatedly to produce new cells. These enter the lymph vessels and are carried into the general circulation.

TISSUE FLUID + LYMPHOCYTES = LYMPH

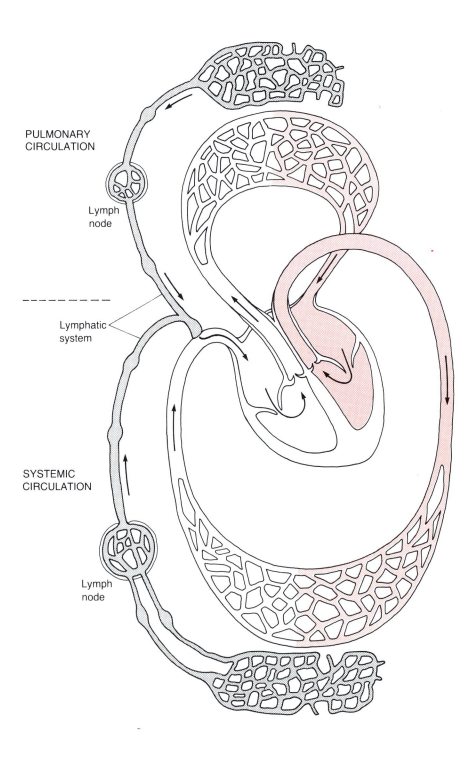

PULMONARY
CIRCULATION

Lymph
node

Lymphatic
system

SYSTEMIC
CIRCULATION

Lymph
node

Fig. 3.4.34   The lymphatic system in relation to the blood system

Fig. 3.4.35 A section through a lymph node

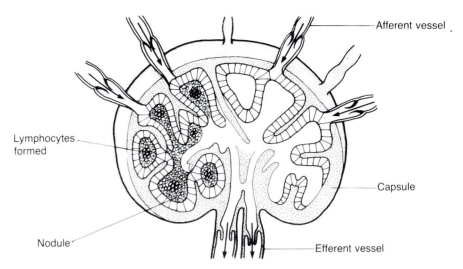

The lymphocytes have similar functions to white blood cells and are able to defend the body against invading bacteria. Valves in the lymph vessels entering and leaving the nodes ensure one-way flow of lymph. As it passes through the nodes it is filtered to remove bacteria. Often lymph nodes occur as groups for example in the groin, armpit and neck. Thus, when an area becomes infected by bacteria, the nodes draining this area may become painful and swollen. Anyone who has a sore throat will notice swollen and painful glands in the neck. This is due to a massive increase in the production of lymphocytes to counteract the bacteria.

The **spleen** is the largest collection of lymph tissue in the body. It filters blood to remove foreign materials and worn out blood cells. The **thymus** is another important lymphocyte producing organ, especially in children. It lies in the chest in front of the heart. After puberty it decreases in size until in most adults it cannot be detected. If the thymus is removed in very early infancy the individual has much less resistance to infection.

Lymph flows much more slowly than blood because there is no heart in the lymphatic system. However, there is some evidence that lymph nodes contract regularly, thus helping to pump lymph around the body. The main flow is maintained in several ways: (a) by the continuous production of tissue fluid; (b) by the contraction of smooth muscle in the walls of the larger lymph vessels; and (c) by the compression of the lymph vessels by surrounding skeletal muscles. Valves ensure a one-way flow back to the vein draining the left arm. In one day the volume of lymph returned to the general circulation equals the total blood volume, about five litres. Thus the lymphatic system is essential in maintaining the volume of circulating blood.

If, for any reason, the flow of lymph back is interrupted, fluid accumulates around the tissues where it causes swelling or oedema. The most extreme form of this is **elephantiasis** in which the lymphatics, usually those in the legs, are blocked by a parasitic worm.

A special route for fat   The inner wall of the small intestine is drawn into folds, the villi (see p. 130), which greatly increase the surface area for absorption. At the centre of each villus is a lymphatic vessel, the **lacteal**, which absorbs fats. Most of the products of digestion are simple, soluble molecules that are absorbed directly into the blood stream. Some products of fat digestion are large and insoluble. These combine in the intestine with bile salts to form small particles (**micells**) which are absorbed by the lacteals. After a fatty meal the lymph contains many of these fat particles and appears milky.

Questions: Domains I and II

1 Copy out the chart below by filling in the blanks to show the composition of blood and its functions.

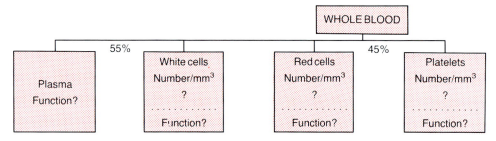

2 Copy out the table below and complete it to show the functions of blood. (I)

Substance carried	From	To	How carried
Oxygen	lungs	body cells via heart	as oxyhaemoglobin
Carbon dioxide	——	——	——
Products of digestion	——	——	——
Urea	——	——	——
Hormones	——	——	——

3 Complete the folowing table by indicating which antibodies occur in the plasma.

Group	Antigen	Antibody
A	A	——
B	B	——
AB	A + B	——
O	None	——

4 A pupil built a model to represent blood circulation. A plan is shown in the diagram.

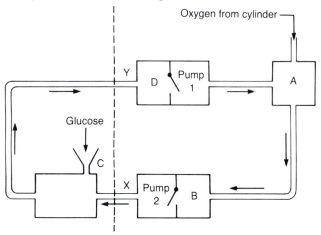

a Compare this with the actual system and name the structures represented by (i) Pump 1; (ii) Pump 2; (iii) A; (iv) B; (v) C; (vi) D. (I)

b State one difference between arteries and veins illustrated by the model. (II)

c The part of the model to the right of the dotted line represents a 'heart–lung' machine, which is sometimes used when a patient undergoes heart surgery. When in use, name the blood vessel which could be attached to: (i) tube Y; (ii) tube X. (I)

d As the blood flows through the machine, oxygen is added. Name a substance which would have to be removed. (I)

5

a List, in the correct order, the blood vessels and chambers of the heart, through which blood passes on its way from the liver to the kidneys. (I)

b Make (i) a table to show the differences between the structures of arteries and veins, and (ii) a table showing the differences in function between arteries and veins. (I)

6

The bar chart represents the average pulse rates of 25 pupils when resting and when engaged in light and vigorous exercises.

a Explain why it is important to measure the *average* pulse rates. (I)

b Give reasons for the differences between the average pulse rates at rest and during exercise. (I)

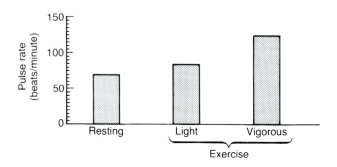

7 Which one of the following diagrams of the right half of a human heart, seen in vertical section, gives the correct position of the valves when the right atrium contracts? (I)

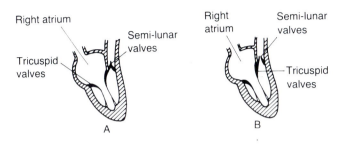

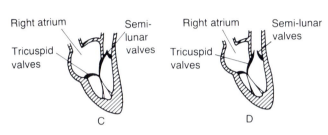

8 The following graph shows the results of red cell counts carried out on a climber training at different altitudes for an expedition in the Himalayas. The count is expressed in terms of red blood cells per dm^3.

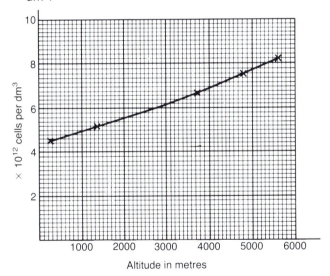

a What was the red blood cell count at:
   i 250m?
   ii 5600m? (II)
b State one factor which could be the cause of the different red cell counts. (I)
c Explain the value to climbers of spending time at high altitudes before undertaking a major climb in the Himalayas. (I)
d Explain one of the effects on the body if this acclimatization is not carried out. (I)
e Suggest why white blood cell counts made during training showed no change. (I)

9 The contraction (systole) and relaxation (diastole) of heart chambers in the heart cycle, which results in one heart beat, can be represented by the diagram shown below where each square represents 0.1 second.

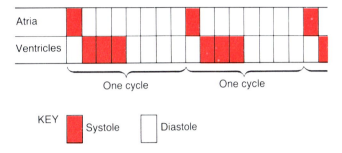

a What is the length of time for:
   i one cycle;
   ii the atrial systole;
   iii the ventricular systole? (II)
b State what happens in the atria during ventricular systole. (I)
c How long does ventricular diastole last? (II)
d How many beats are made in one minute? (II)

10 The graph shows pressure changes in the left ventricle and aorta during the beating of the heart.

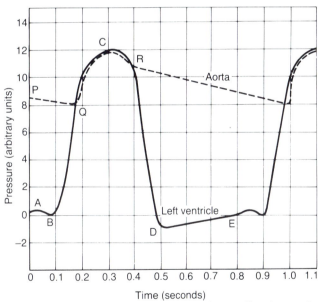

a Use the letters on the graph and describe the parts representing
   i contraction of the left ventricle;
   ii closure of the semi-lunar valves;
   iii relaxation of the left ventricle;
   iv closure of the bicuspid (mitral) valve. (II)
b State the highest pressure measured in the left ventricle. (II)
c Explain why the pressure in the aorta does not vary as much as that in the left ventricle. (I)

Experimental skills and investigations: Domain III

1 Measuring the rate of heart beat

Materials
Rubber tubing, glass T-piece, filter funnel, stop clock.

Procedure
1 Construct a stethoscope as shown in the diagram.
2 Place the stethoscope in your ears and put the end of it to the left of centre of your partner's chest.

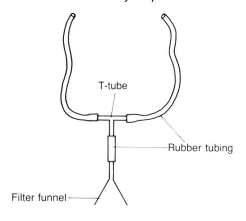

3 Count the number of heart beats per minute at rest.
4 Describe the sounds you heard.
5 From your observations and your knowledge of the way in which the heart works, suggest what the sounds represent.

2 To find out the direction of flow of blood

Materials
A piece of cloth large enough to be tied around a person's arm.

Procedure
1 Tie the cloth (not too tightly) around the person's arm.
2 Get the person to make the veins in the arm visible by clenching the fist.
3 Press a finger against a large vein. Slide this finger, with a light pressure, a few cm along the vein towards the wrist and continue to hold your finger on the vein.
4 Repeat stage 3, sliding your finger towards the elbow.

OBSERVATIONS
Record
1 The appearance of the vein after sliding your finger towards the wrist.
2 The difference made by changing the direction of sliding the finger.

From your observations, deduce the direction of blood flow.

3 To find out which blood groups can be mixed together safely

Materials
A white tile, a glass tube, eight stock solutions representing blood groups. These are:

Donor groups
O – Distilled water and eosin
A – $FeCl_3$ and eosin
B – K_2SO_4 and eosin
AB – $FeCl_3$ plus K_2SO_4 plus eosin

Recipient groups
O – $BaCl_2$ plus NaOH plus eosin
A – $BaCl_2$ plus eosin
B – NaOH and eosin
AB – Distilled water and eosin.

Procedure
Using the glass tube, spot samples of the artificial blood on to the tile according to the following instructions. WASH THE TUBE CAREFULLY BETWEEN EACH SAMPLE WITHDRAWN OR THE STOCK SOLUTIONS WILL BECOME CONTAMINATED AND USELESS.

Mix the following solutions on the tile:
1 Donor A + Recipient A
2 Donor A + Recipient B
3 Donor A + Recipient AB
4 Donor A + Recipient O

In each case record whether a clot is formed (shown by a white or brown solid in the red liquid), or whether they mix without clotting.
   Wash the tile thoroughly and begin the investigation again using Donor B instead of Donor A. Record the results as before. Wash everything thoroughly again. Repeat using Donor AB and Donor O.
   Make a table of results as shown below:

	Donor O	Donor A	Donor B	Donor AB
Recipient O				
Recipient A				
Recipient B				
Recipient AB				

From your observations deduce
1 Which donor group can mix with any other group.
2 Which recipient group can mix with any other.
3 Why it is important that the tile should be washed thoroughly.

4 To find out the effect of exercise on pulse rate

Materials
Stop clock, graph paper.

Procedure
1 Sit quietly. Using your finger tip, find your pulse where an artery passes over a bone near to the surface of the skin in your wrist, as illustrated in the diagram. Do not use your thumb because you have a pulse in it which might confuse your results.
2 Count the number of times your pulse beats in 15 seconds.
3 Write down the result at 1 below.
4 Count your pulse twice more over 15 seconds and record the results at 2 and 3 below.

OBSERVATIONS
1 The first count of my resting pulse was _____ beats in 15 seconds.
2 The second count of my resting pulse was _____ beats in 15 seconds.
3 The third count of my resting pulse was _____ beats in 15 seconds.
4 Calculate your average resting pulse rate by adding all three results together and dividing your total by three.

5 Calculate your average resting pulse rate per minute by multiplying your last answer by four.
6 IF THERE IS ANY MEDICAL REASON WHY YOU SHOULD NOT EXERCISE – TELL YOUR TEACHER NOW.
7 Exercise for five minutes by press-ups or Boston steps. Really put yourself to the test but not for more than five minutes.
8 Sit down and find your pulse again.
9 Record your pulse every 15 seconds until it returns to normal.
10 Convert your results to pulse rates per minute by multiplying them all by four.

Record your results in a table similar to the following.

Time after exercise every 15 seconds	Pulse counts	Pulse rate per minute
0–15		
15–30		
30–45		
45–60		
60–75		
75–90		
90–105		
105–120		
120–135		
135–150		
150–165		
165–180		
180–195		
195–210		
210–225		

11 Use graph paper to plot a graph of your pulse rate per minute against time after exercise in seconds.

From your observations deduce
1 Your resting pulse rate per minute.
2 The length of time taken for your pulse to return to normal.

Suggest why your pulse rate did not return to normal as soon as you stopped exercising.

NOTE
Do not pressurise students into doing physiological exercise of this kind if they are not healthy or fit enough to do it.

3.5 Organs that hold your breath

The breathing system

The function of breathing is to provide oxygen to the body's cells and to remove carbon dioxide from these sites. Simple organisms obtain the necessary oxygen and eliminate carbon dioxide by diffusion across their body surfaces. An organism cannot be larger than about 0.5mm in diameter to survive by this means alone. Larger organisms have developed respiratory systems that enable them to overcome this problem. In Man, the lungs bring large volumes of gas from the atmosphere into close contact with blood in the circulatory system. Blood can carry enough oxygen to all parts of the body.

The necessary gas exchange takes place within the lungs (see Fig. 3.5.1). These organs provide, in a relatively small volume, a very large surface area. Membranes separating air and blood are very thin and always moist. Across this surface, molecules of oxygen and carbon dioxide pass in solution by simple diffusion.

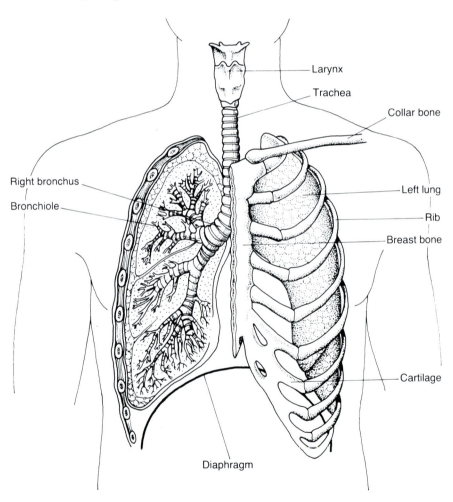

Fig. 3.5.1 The breathing organs

The body's air conditioner

Air enters the lungs through the upper air passages – the nose and mouth. It passes through the **pharynx**, **larynx** and **trachea** or **windpipe**. These passages are lined with epithelium, containing mucus-secreting cells. They produce a thin overlying mucous blanket, and have a layer of hair-like **cilia** (see Fig. 3.5.2). Beneath this epithelium there is a rich network of capillaries.

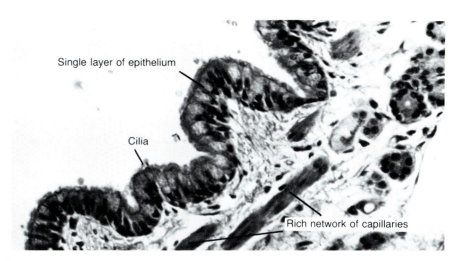

Single layer of epithelium

Cilia

Rich network of capillaries

Fig. 3.5.2 Photomicrograph of ciliated epithelium

The function of this upper part of the respiratory system is to act as a filter. It traps harmful particles on the lining of these passages. Once such material has been deposited it is removed. This is done by the upward movement of the mucous blanket caused by the coordinated beating of the underlying cilia. The cough mechanism and sneezing can both aid this process. This part of the respiratory system humidifies and warms the cool dry air, which is inhaled, with its rich blood supply. This prevents drying of the delicate gas-exchanging membranes deep within the lungs.

Branching like a tree

Soon after entering the chest, or **thorax**, the trachea divides into right and left main **bronchi**. Each bronchus divides again into several branches. Within the lung, each of these bronchi subdivides into two again and again, some twenty times. At each division the size of the bronchi decreases. Twenty divisions of this type produce about one million

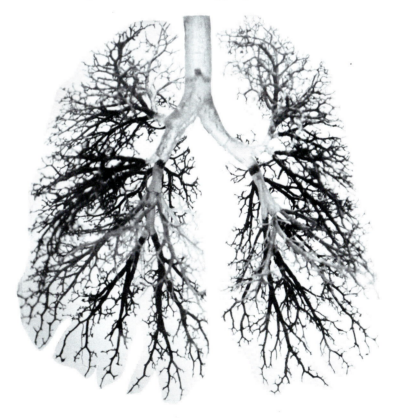

Fig. 3.5.3 The respiratory tree

Fig. 3.5.4 Cartilage supports of the trachea and bronchi

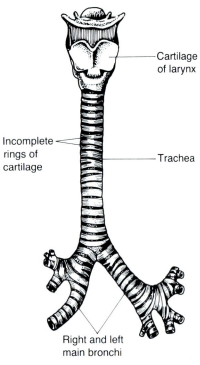

terminal branches called **bronchioles**. Like the trachea, the larger bronchial branches are lined with cilia and a mucous blanket. In the main bronchi and trachea, cartilage forms regularly-spaced horseshoe-shaped supports (see Fig. 3.5.4). The supports are necessary because of the pressure differences within the system. The bronchi and bronchioles have relatively thick walls. They do not allow the exchange of gases to take place. The finest of their branches divide to form the first part of the true respiratory tissue of the lungs. They lead to the air sacs called **alveoli** (see Fig. 3.5.5).

It is here in the alveoli that gas exchange occurs (see Figs 3.5.6 and 3.5.7). There are about three hundred million of these alveoli in human lungs. Each is in very close contact with a rich network of capillaries. There is a vast and extremely thin surface for the transfer of gases between air and blood. The surface area of this membrane is approximately seventy square metres, or forty times the surface area of the body, and less than one thousandth of a millimetre thick.

Fig. 3.5.6   Close up of an alveolus

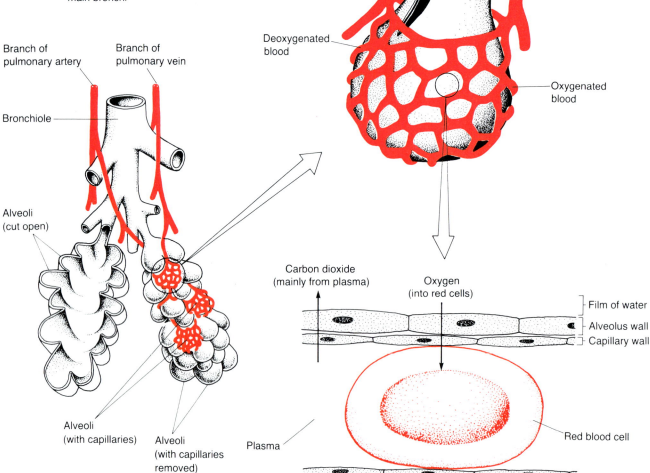

Fig. 3.5.5   Alveoli at the end of a bronchiole

Fig. 3.5.7   How gases pass in and out of an alveolus

An elastic sponge

Fresh air is drawn into the lungs by increasing the volume of the thorax and making its pressure less than that outside. The thoracic cavity is enlarged by contracting the **muscles of inspiration** (see Fig. 3.5.8). The chief one is the **diaphragm**. It is dome-shaped, separating the thorax from the abdomen. On contraction it increases the thoracic volume. It moves downward causing the rib cage to expand. This motion is aided by the contraction of the muscles between the ribs, the external intercostal muscles.

Expiration is usually a passive process and is achieved by relaxation of the inspiratory muscles (see Fig. 3.5.8). The amount of air the lungs can hold is called their **vital capacity**. The air which is left inside the lungs after breathing out as far as possible is the **residual volume**. These are shown in Fig. 3.5.9. When air is drawn into the lungs they act like an elastic sponge. It is the elastic recoil of this sponge that provides the forces required to expel air during quiet expiration.

Inspired air at sea level contains 21 per cent oxygen but only 0.04 per cent carbon dioxide. Air in the alveoli, however, contains slightly less oxygen (14 per cent) but considerably more carbon dioxide (5.5 per cent). As a result, oxygen from the alveoli diffuses through the lining, through the capillaries and into the plasma. The final stage of the diffusion path is from the plasma to the red blood cells. Carbon dioxide diffuses along the reverse pathway. The process is rapid. By the time a red cell has moved around the surrounding capillary, it is complete.

The amount of oxygen with which the haemoglobin can combine depends on the partial pressure (proportion) of the oxygen in the alveolus. Normally the pressure is so high that the red cells combine fully with the oxygen. The haemoglobin of these cells is then bright red. On reaching the body tissues, about one third of the oxygen is given up. This is caused by the low pressure of oxygen in active tissue. The haemoglobin is then 60 to 70 per cent saturated and is returned to the heart. Haemoglobin with a low oxygen content is bluish in colour.

Fig. 3.5.8   The mechanism of breathing

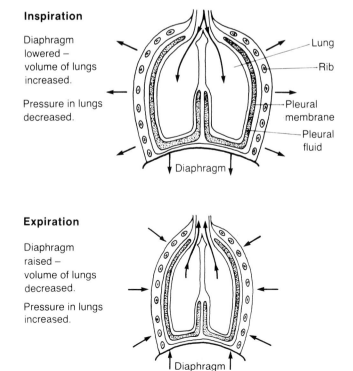

Inspiration

Diaphragm lowered – volume of lungs increased.

Pressure in lungs decreased.

Lung
Rib
Pleural membrane
Pleural fluid

Diaphragm

Expiration

Diaphragm raised – volume of lungs decreased.

Pressure in lungs increased.

Diaphragm

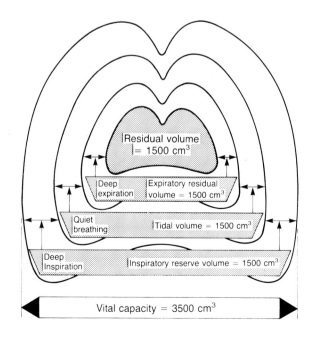

Residual volume = 1500 cm³

Deep expiration

Expiratory residual volume = 1500 cm³

Quiet breathing

Tidal volume = 1500 cm³

Deep Inspiration

Inspiratory reserve volume = 1500 cm³

Vital capacity = 3500 cm³

Fig. 3.5.9   Subdivisions of lung air

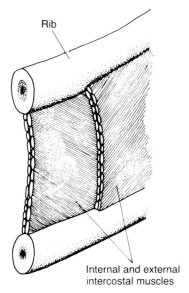

Fig. 3.5.10 Muscles involved in breathing – intercostal muscles

Automatic control

The breathing mechanism is regulated and controlled by the **central nervous system**. It relies on the rhythmic contraction and relaxation of the respiratory muscles (see Figs 3.5.10 and 3.5.11). Under normal circumstances a person is unaware of this process. It continues during sleep and needs no conscious effort. The control of the rhythmic contractions originates in the respiratory centre of the hind brain. Impulses from the centre pass via nerves to the diaphragm and the intercostal muscles. As a result, inflation and deflation of the lungs occur.

The respiratory centre is sensitive to the levels of oxygen and carbon dioxide as well as the acidity of the blood (see Fig. 3.5.12). The body is most sensitive to the carbon dioxide in the blood. Its concentration is sensed by specialized receptors close to the respiratory centre of the brain. A high level stimulates an increase in both rate and depth of respiration.

The respiratory system is one of the most complex in the body. It has to bring the blood near to the air in the lungs. These organs are unusual in bringing the internal and external environments close together. Exchanges between them can then occur. At rest the lungs take up about $200cm^3$ of oxygen per minute. It can be increased ten times on exercise. Because of the control mechanisms, this increase occurs within a few seconds.

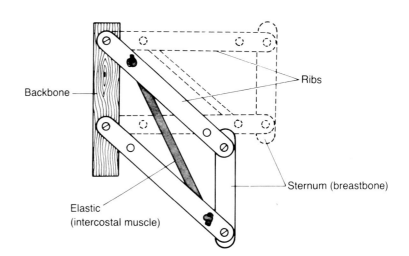

Fig. 3.5.11 A model to show the action of the muscles involved in breathing

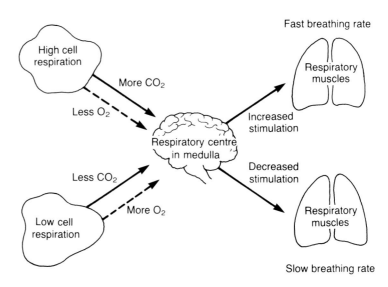

Fig. 3.5.12 Automatic control of breathing

Life's energy

Respiration is the release of energy from glucose in every living cell of a living organism. Energy is the 'push' that makes things happen. It is the ability to do work and can be classified in two ways. If you wind up the spring of a clock, the tightly wound spring has stored or potential energy in it. As the spring begins to unwind and turn wheels, the moving, active energy is called **kinetic energy**.

When carbon atoms are joined together, their links or bonds are a form of potential energy. The bonds are similar to wound-up springs. A log burning in a fireplace is undergoing a chemical change. Stored energy in chemical bonds is being released. The carbohydrate in the log breaks down into carbon dioxide and water. The energy in the bonds is rapidly converted to heat and light – the flames. Humans also give off carbon dioxide and water and the process is similar to burning. Although the end products of the two processes are the same, cells 'burn' their fuel in quite a different way. Instead of letting the energy escape rapidly, cells must save it to be used for many purposes. Chemical reactions in cells take place in small steps that release energy slowly.

There is transfer of energy in every chemical reaction. Some reactions release energy and others store it up. In cells these two types of reactions are linked by special chemical compounds that store energy and then release it. The most important one of these contains three groups of phosphate ions (a combination of phosphorus and oxygen). It is called **adenosine triphosphate** or **ATP**. This can be represented as follows:

$$A - P - P - P$$

where A = adenosine and P = phosphate.

The three phosphate groups form a chain, hooked to the rest of the molecule. The loss of one phosphate ion changes ATP into **adenosine diphosphate**, **ADP**.

$$A - P - P - P \ = \ A - P - P \ + \ P$$

The last phosphate group is easily separated, together with its bond, from the rest of the ATP molecule and joined to other compounds.

A molecule that receives the phosphate group from ATP can either use the energy for some chemical reaction, for example, joining amino acids together or making muscles contract, or it can pass it on again. Sooner or later, however, the energy in the phosphate bond will be used and the phosphate group will be released. The free phosphate group will then join to an ADP molecule and recycle ATP.

The ATP in animal cells is used to split glucose molecules to obtain their potential energy for use in the cells' chemical reactions.

The chemical changes that take place in respiration are simplified in the equation:

$$C_6H_{12}O_6 \ + \ 6O_2 \ = \ 6CO_2 \ + \ 6H_2O \ + \ E$$

| Glucose | Oxygen | Carbon dioxide | Water | Energy |

Some diseases of the breathing system

If the air we breathe contains irritating particles and gases, it is not surprising that the breathing system can suffer. Two of the most common diseases caused by such irritants are **bronchitis** and **emphysema**.

Bronchitis is the inflammation and bacterial infection of the bronchi. Both environmental air pollution and self-induced air pollution by smoking, can be important influences in its development. On top of these factors is the added burden of infection. It can strike anyone at any time, but may be particularly severe in winter.

Inflammation of the bronchi usually begins as an infection by a virus, followed by bacterial infection. The larynx and trachea are often also affected causing **laryngitis** and **tracheitis**. Bronchitis may spread to the lungs causing **pneumonia**. The first symptoms are a dry cough and a feeling of soreness in the centre of the chest. There may be symptoms of a cold and also some hoarseness in the throat. After two or three days, yellowish-green mucus with pus is coughed up and there will be a moderate fever. Patients with previously healthy lungs recover naturally from mild bronchitis in about two weeks. Heavy smokers may continue to cough for six weeks. Severe cases may be treated with the antibiotics tetracycline and ampicillin. A medicine containing codeine is useful at any stage to ease the cough.

Emphysema is a condition in which the lungs are not properly emptied between breaths. It almost invariably occurs with severe bronchitis, narrowing the airways and causing wheezing. It greatly aggravates the shortness of breath. A serious complication which can arise is the accumulation of air in the cavities around the lungs. It is caused by a breakdown of lung tissue, allowing air to escape into the pleural cavity.

Dying for a cigarette – smoking

Every year in the United Kingdom about 100 000 people die through illness caused by smoking. Millions of pounds are spent on improving the safety of motor-cars to reduce road accidents. Strict laws try to reduce the death toll on roads due to drunken drivers. Yet seven times as many people are killed as a result of smoking and far less money is spent on reducing this death toll.

Facts

Consider a school population of 1000 pupils. Assume that all of them smoked. If we consider national statistics, it is quite likely that of the 1000 pupils one will be murdered, six will die in road traffic accidents and 250 will die early as a result of smoking-related diseases.

Ninety per cent of all lung cancers are caused by smoking. A person who smokes five cigarettes a day is six times more likely to die of lung cancer than a non-smoker. A person who smokes 20 a day increases the risk to 19 times more than for a non-smoker.

Tobacco also causes cancer of the mouth, throat and wind-pipe. Ninety-five per cent of all people who suffer form bronchitis are smokers. Smokers are two to three times more likely to die of heart disease than non-smokers.

Babies born to women who smoke during pregnancy are on average 200g lighter than non-smokers' babies. A mother who smokes 20 a day during pregnancy increases the risk of the baby dying at birth by 35 per cent and increases her risk of bleeding seriously during pregnancy and producing less milk.

Fig. 3.5.13   A disgusting mess

An unsociable habit

People who share a house with a smoker are also likely to suffer from smoke-related diseases. Children with parents who smoke are at risk from bronchitis and pneumonia. An hour in a smoke-filled room leads non-smokers to inhale as much harmful material as if they had smoked 15 filter-tipped cigarettes. Non-smokers' eyes, noses and throats are also irritated by smoke.

Fig. 3.5.14 Discharge from an average smoker's lungs

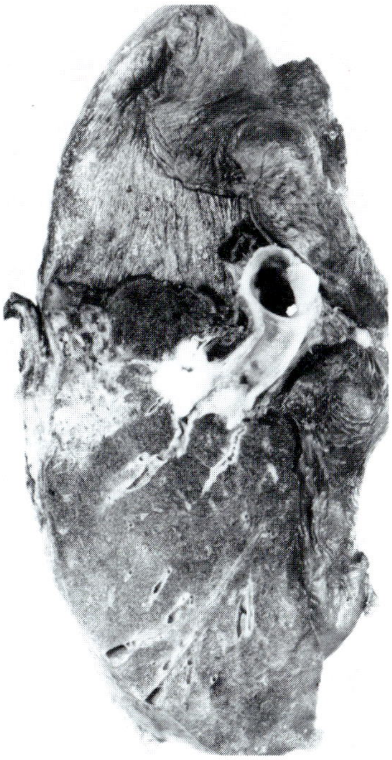

Fig. 3.5.15 Photograph of a lung showing carcinoma (cancer)

Paying to die

Between 1980 and 1981 the Government collected £3350 million in taxes from smokers. The Health Service spends about £2 million a week on treatment and in addition the Government pays out sickness benefit, widows' pensions and social security benefits to the families of victims of the smoking habit.

Fifty million working days per year are lost by industry because of ill-health caused by smoking. About 20 per cent of all factory fires are caused by discarded cigarette ends.

So, why do people smoke? Simply because they can't stop once they have started. Tobacco contains nicotine which causes addiction.

A much more difficult question to answer today is 'Why do people start to smoke?' The tobacco industry makes a few people very rich and kills a great many others. The art of advertising has reached such a degree of perfection that it achieves its aim without people really knowing that they are being 'sold' products. Vast sums of money are spent on advertising the desirable image of smoking; and it works. The Health Education Council tries to counteract this image but cannot compete with the sums of money spent on advertising by the tobacco industry.

The deadly substances in cigarettes include:

1 **Tar**. This is the main cancer-causing substance.
2 **Nicotine**. This is an addictive drug.
3 **Carbon compounds**. These can be changed to the poisonous gas, carbon monoxide.
4 **Acreolin**. An irritant substance which causes coughing.

Questions: Domains I and II

1 The table shows the tar and nicotine content of cigarettes.

Brand of cigarette	Plain (P) or Filter (F)	Tar yield (mg/cigarette)	Nicotine (mg/cigarette)
Bristol	F	8	0.4
Kool	F	15	1.0
Players No. 3	P	33	2.1
Guards	F	20	1.3
John Player Special	F	22	1.4
Capstan Full Strength	P	38	3.2
Silk Cut Extra Mild	F	4	less than 0.3

a   Which brand has the lowest tar content? (II)
b   What effect does the filter have on:
   i   the tar content;
  ii   the nicotine content? (II)
c   What is the average tar yield per cigarette for the seven brands? (II)
d   What effect does nicotine have on the cilia found in the trachea? (I)
e   Name two lung disorders that are more likely if a person smokes regularly. (I)

2 The diagram shows an apparatus which may be used to show the mechanism of breathing.

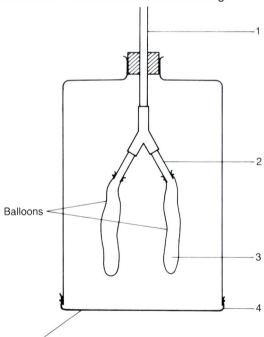

a   If the rubber sheet is pulled down, air enters the balloons making them inflate slightly. Describe what happens following the pulling down of the sheet to cause this change. (I)
b   The apparatus can be thought of as a model of the thorax. What do the numbered structures represent in the body? (I)

3 The graph shows the intake of air during rest, then during a period of 10 minutes vigorous running and finally during rest following the running.

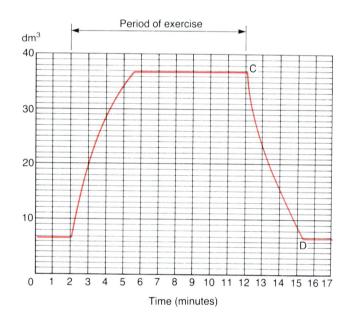

a   What was the volume of air breathed in each minute during the first period of rest? (II)
b   What is the maximum volume of air breathed in each minute during exercise? (II)
c   Give the reason for the shape of the graph between points C and D. (I)
d   It has been found that muscle, which is not contracting, uses about 0.15cm^3 oxygen per 100g of muscle each minute. Muscle which is actively contracting such as when running fast, uses about 12cm^3 oxygen per 100g of muscle each minute.
   i   For what purpose is oxygen used in all living cells? (I)
  ii   Explain the reason for the difference in the volumes of oxygen used at rest and during exercise. (I)
  iii   Give two ways by which the body is able to increase its oxygen intake during exercise. (I)

Experimental skills and investigations: Domain III

1 How does exercise affect your rate of breathing?

Procedure
1 While resting, place your hands on the lower part of your rib cage and count the number of inspirations over a period of thirty seconds. For this to be valid you must breathe naturally – do not force your breathing. Repeat this four times so that you can calculate your average rate of breathing at rest. Record the average reading and double it to find your breathing rate per minute.
2 Run on the spot vigorously for one minute, stop and repeat the first procedure. Again, repeat four times to find an average reading for your rate of breathing after exercise.
3 Draw a bar chart to show the change in breathing rates of the pupils in the class after exercise.

Questions
a What is your average breathing rate before and after exercise?
b How close is your average to the average for the class?

2 Is there as much carbon dioxide in expired air as there is in inspired air?

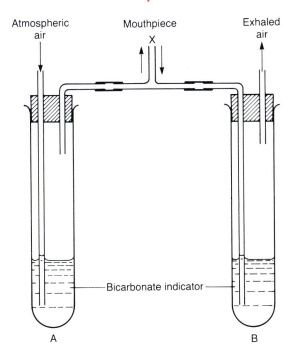

Procedure
1 Set up the apparatus as shown in the diagram. Pay particular attention to the precise arrangement of the tubes.

2 Breathe in and out at point X about ten times and record any changes in colour of the indicator in tubes A and B.

Questions
a Describe the nature of any colour change observed.
b Explain why the contents of one of the tubes changed colour before the other.

3 Is expired air warmer than inspired air?

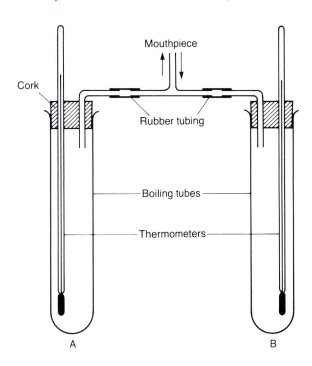

Procedure
1 Set up the apparatus as shown in the diagram. The corks must be secure and the bulbs of the thermometers must not touch the sides of the tubes.
2 Record the temperature in each boiling tube.
3 Pinch the rubber tubing next to tube B so that its entry is closed and draw in air through A. Do this by gently drawing in air through the mouth piece, about ten times. Record the temperature in boiling tube A.
4 Close the entry to tube A, in a similar way, then breathe out gently through B, about ten times. Record the temperature in boiling tube B.

QUESTIONS
a State your conclusion.
b Why is the apparatus supported by retort stands rather than by your hands?

4 How much air do you breathe in during normal inspiration?

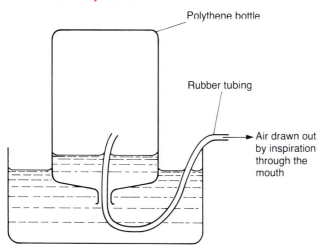

Procedure

1 Put a length of rubber tubing through the neck of a large container (a 5dm³ polythene bottle is ideal) until it nearly reaches the bottom.
2 Invert the bottle and place it under water in a laboratory sink (see diagram).
3 Hold your nose and practise breathing in and out through your mouth.
4 Breathe in air from the inverted bottle. Repeat four times.
5 After breathing in, quickly remove the rubber tubing, insert a stopper in the bottle neck and remove the bottle.
6 Measure the volume of water contained in the bottle using a measuring cylinder.
7 Repeat several times and take the average volume.

QUESTIONS
a State the volume of water.
b What does this volume represent?

5 How much can you breathe out in a single breath?

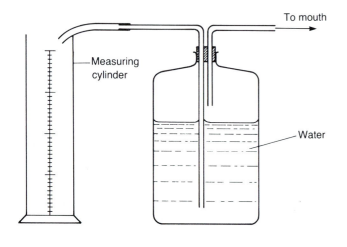

Procedure

1 Set up the apparatus as shown in the diagram.
2 Practise deep breathing a few times, then blow out deeply through the mouth into the apparatus.
3 Collect the displaced water in the largest available container (at least 5dm³).
4 Measure the volume of water collected. Use a measuring cylinder.
5 Repeat several times and take your highest reading. This is your vital capacity.

QUESTIONS
a What is your vital capacity?
b Draw a bar chart for the vital capacities of all the pupils in the class.
c Compare this with the bar chart made for the breathing rates and exercise.

6 What parts of cigarette smoke stays in your lungs?

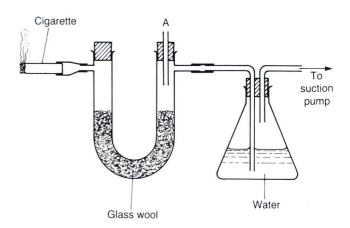

Procedure

1 Set up the apparatus as shown in the diagram.
2 Set up a second apparatus without a cigarette.
3 Turn on the suction pump and place a finger periodically over A, causing air to be drawn through the cigarette.

QUESTIONS
a What is the purpose of the apparatus without the cigarette?
b Describe the appearance of the glass wool at the end of the experiment in both sets of apparatus.
c Describe how you would use the apparatus to compare the levels of tar obtained from plain and filter-tipped cigarettes.

3.6 Plumbing and heating

Kidneys – cleansers of the blood

The **kidneys** are a pair of organs each having a tube called a **ureter** (see Fig.3.6.1). The ureters carry **urine** to the **bladder**, where it is stored before being eliminated through the **urethra**. The kidneys are situated on each side of the spine in the posterior of the abdomen. They are well-protected, lying under cover of the diaphragm and embedded in fat. Two **renal arteries** supply them with blood from the aorta. The rate of blood flow through these arteries is about 1.5 litres each minute. This enormous blood supply to such a small part of the body highlights the importance of the kidneys.

The funnel-shaped upper-end of each ureter is called the **pelvis**. This divides within the kidney to form cup-like formations for collecting urine.

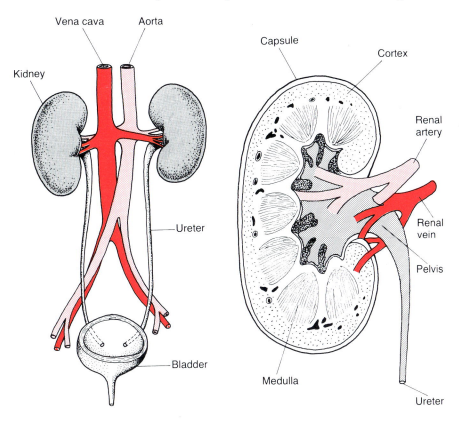

Fig. 3.6.1 The urinary system

Fig. 3.6.2 A section through a kidney

A complex pipeline

Each kidney is covered by a thin transparent **capsule** that can be peeled away to show the **cortex** (see Fig.3.6.2). This surrounds the paler internal area known as the **medulla**. The medulla is built-up of cone-like masses of tissue called **pyramids**. The fine structure of the kidney can only be studied with the aid of a microscope. It is built-up of an extremely large number of pipes called **tubules**. Each consists of a **nephron** or secreting part which produces urine, and a collecting tubule (see Fig.3.6.3). The nephron is made of a filter called a **Malpighian corpuscle**, and the **renal tubule**. The latter is concerned with reabsorption of material passing through it. There are up to one million Malpighian corpuscles per kidney. They are made of (1) a central **glomerulus** of tiny capillaries and (2) a surrounding membrane called the **Bowman's capsule**.

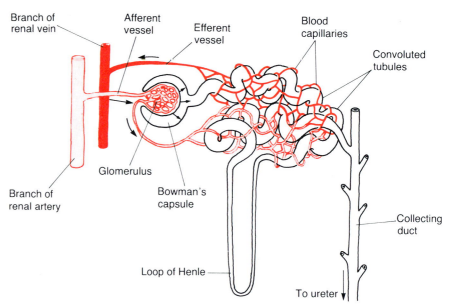

Fig. 3.6.3 A single nephron with its blood supply

The network of tiny capillaries is derived from a small artery called the **afferent vessel**, which enters the glomerulus, and the corresponding **efferent vessel** which leaves it at the same site. In order to increase the rate of filtration under pressure, the diameter of the afferent vessel is larger than that of the efferent.

The renal tubule consists of several divisions. Beginning at the capsule end, the first part is the **proximal convoluted tubule**. This runs into the descending limb of the **loop of Henle**. The latter is sited in the medulla. It leads to the **distal convoluted tubule**. The proximal tubules are about 14mm long and lined with pyramid-shaped cells. The inner surface of these tubes is covered with microvilli, which increase the surface area for reabsorption (see Fig.3.6.4).

The loop of Henle and the distal tubule are built up of different shaped cells, and the diameter of this pipeline becomes narrower as it progresses towards the collecting system. The length of a single nephron is approximately 50mm. The total length of all the tubules from both kidneys joined end to end would be 110 kilometres. Some cells of the kidney produce an enzyme called **renin** which increases the blood pressure in the Malpighian corpuscles. The effect of this is to increase the rate of filtration through the Bowman's capsules under pressure.

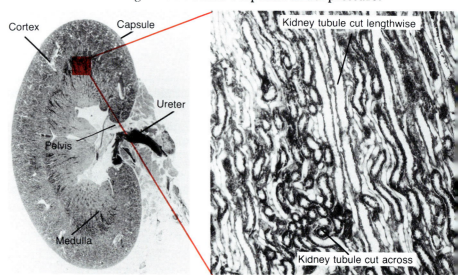

Fig. 3.6.4 Photomicrograph of a transverse section through a kidney tubule

Water regulation

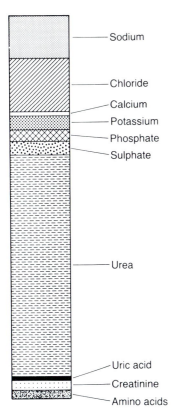

— Sodium

— Chloride

— Calcium

— Potassium

— Phosphate

— Sulphate

— Urea

— Uric acid

— Creatinine

— Amino acids

Fig. 3.6.5 The average composition of urine

The basic principles of kidney function are (1) that plasma is filtered in the glomeruli and (2) that the concentration of sugar, amino acids, salts and water is regulated by reabsorption. Each kidney receives a blood flow of between 500 and 700cm^3 each minute. In the glomerulus 120cm^3 of fluid is filtered for this and passed into the Bowman's capsules. The filtrate is rich in glucose, sodium, potassium, calcium, phosphates and chlorides. It also contains urea and a large volume of water. Each part of the renal tubule is responsible for the reabsorption of a particular constituent of the filtrate.

The distal tubule is the site at which water balance is adjusted. Regulation is controlled by a chemical messenger called **ADH** or **antidiuretic hormone**. It is one of the hormones made by the pituitary gland in the brain.

When water intake is low, very little is absorbed through the wall of the small intestine to the blood. Consequently, the amount of water in plasma is decreased and the blood becomes 'more concentrated'. It is this highly concentrated blood, circulating through the pituitary gland, that triggers a response. The pituitary gland secretes lots of ADH. This circulates in the blood and eventually reaches the kidneys. Here, the membranes of the tubule cells become affected. They become more permeable and a lot of water is reabsorbed so that the level of water in the body stays constant. Small quantities of highly concentrated urea are excreted.

When the water intake is high, a large volume of water is absorbed through the wall of the small intestine to the blood. Consequently the amount of water in the plasma is increased and the blood is 'diluted'. This 'diluted' blood does not cause the pituitary to produce ADH. The membranes of the tubule cells are not so permeable and little water is absorbed. Large quantities of dilute urea are excreted. Fig. 3.6.5 shows the normal composition of urine.

The artificial kidney

Kidney machines can be used to purify blood when the kidneys become damaged or diseased and are unable to carry out their normal functions (see Fig. 3.6.6.). The principle of the machine is to take a person's blood, pass it through a filter, and put it back into the person after waste materials have been removed (see Fig. 3.6.7).

In many cases a permanent 'shunt' is put between an artery and vein of the arm. The blood is then led from the artery to the artificial kidney and then returned to the vein.

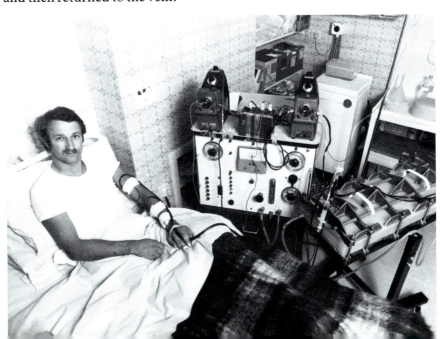

Fig. 3.6.6 A kidney machine in use

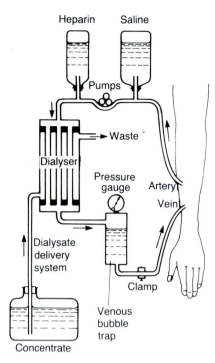

Fig. 3.6.7   The principle of a kidney machine

The machine has a **selectively permeable membrane**, like visking tubing, to filter the blood. The whole apparatus is placed in a tank containing a fluid which takes the waste away after filtration. The fluid has a concentration of salts which is the same as that of the blood. The membrane allows water, urea, uric acid, ammonia and salts to pass through but will hold back larger molecules of glucose, fatty acids, amino acids, and proteins.

Sodium and potassium will pass freely into the fluid. These salts are essential to the body so their levels are kept constant. It is difficult to prevent the movement of these salts out of the blood into the fluid. That is why the fluid contains the same concentration of these salts as the blood. These salts can pass both ways across the membrane.

It is possible for people to use kidney machines in their homes. The process of filtration takes from three to six hours and is repeated three times per week.

Another type of treatment is now available to people suffering from kidney failure. It uses the body's own abdominal lining, the **peritoneum**, as a selectively permeable membrane. A fluid with the same salt concentration as blood is introduced into the abdomen through a tube (a catheter). The tube remains permanently in place and is capped when not in use. The peritoneum acts as a filter between the body fluids and the blood supply of the alimentary canal. About two litres of fluid have to be replaced four times a day with fresh fluid. The process takes about thirty minutes and can be done almost anywhere.

What others see of us – skin
(see Fig. 3.6.8)

The human skin is a two-layered organ. It consists of (1) a deeper layer called the **dermis**, fibrous and supplied with blood vessels and nerves and (2) the **epidermis** consisting of bloodless layers of cells (see Figs. 3.6.9 and 3.6.10). The epidermis depends on the dermis for its nourishment.

(a)

(b)

Fig. 3.6.8   (a) The smooth elastic skin of a baby's face (b) The wrinkles of old age

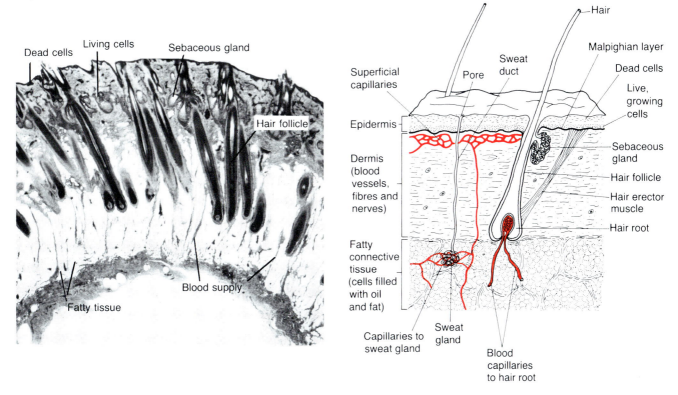

Fig. 3.6.9 Photomicrograph of a section through the skin   **Fig. 3.6.10** A section through the skin

These two layers of skin lie on another layer of fatty connective tissue. This provides some of the protective covering of the body. It is the place where fat is stored as a source of energy and it acts as an insulating layer. The skin varies greatly in structure from one region to another. There are, however, only two main types of human skin: (1) hair-bearing, which occurs over most of the body and (2) smooth, which is present on the palms and soles. Smooth skin has a thick outer layer, designed to stand up to buffeting and pressure. It has sweat glands but no hair or grease glands, thus providing a powerful grip at places where this is most needed. At the same time it is very well supplied with nerve endings.

The dermis consists of strands of protein and elastic fibres called **collagen**. These enable the skin to recover its shape after it is stretched. Throughout life this elasticity gets less until it gives way to the sagging skin of old age. Among the fibres run numerous blood vessels. They are under automatic nervous control which opens or shuts the flow of blood in the outermost blood vessels. This provides a fine adjustment for temperature regulation. The dermis is well supplied with sensory nerves that recognize touch and pain. These vary in number from one part of the skin to another. Sensitivity is greatest on the finger tips and much less at other places such as the back.

The epidermis develops from the cells in a specially active layer called the **Malpighian layer**. As the cells divide they gradually move to the surface. They lose their nuclei and become the pavement-like flakes of dandruff. A protein, **keratin**, makes up the bulk of these dead cells. It is soft on the surface of the skin but hard when in hair and fingernails.

Scattered among the Malpighian layer are the pigment-forming cells, **melanocytes**. These make **melanin** when stimulated by sunlight and give the skin a dark appearance. Other melanocytes in the hair follicles give colour to the hair. On the outer part of the epidermis, the melanin acts as a sun screen.

A natural cosmetic In most places the epidermis is perforated by hairs and the openings of **sweat ducts**. Each hair has a **sebaceous gland** which makes an oil, **sebum**. It softens and protects the cells beneath. The **sweat glands** lie in the dermis. They are coiled and run to the surface as sweat ducts. There are over 400 per square centimetre on the palms but only about eighty per square centimetre on the thighs. Each hair has an **erector muscle** also under automatic nervous control. By contracting, it pulls the hair to a more upright position. This is a way of insulating mammals but for us clothes are more important.

Our central heating system

Heat is continually being produced by the body. In the resting state, some of this heat must be lost if the body temperature is not to rise. The heat arises because the body is not very good in the way it uses energy from food. Ninety per cent of the energy from our daily intake of food appears as heat within the body. Most of this is released in the liver or muscles.

The body is like a car engine. A car also produces heat as a waste product in its engine. Like a car, the body must have some means of getting rid of the excess heat.

The temperature of all mammals and birds is kept constant. Such animals are called **homeotherms**. The advantage of being one is that the animal can live in a great variety of climates.

Transfer of heat The temperature of the body is not the same all over. Heat production in some organs is greater than in others. Much of the heat is transferred to the skin in the blood. Thus the circulation of blood, like that of water in a car engine, carries heat from the inside to the outside. The circulatory system also helps to even out the temperature differences within the body. Warm blood, draining an active muscle, is carried into the general circulation.

The most common site at which a person's temperature is taken is the mouth. The thermometer is placed under the tongue and the mouth kept closed for at least two minutes. Even longer is needed if the person has been breathing cold air or drinking hot liquids. Temperatures recorded from the mouth are usually lower than those recorded from the rectum. Rectal temperatures are often taken if a person is delirious or recovering from a mouth operation. Alternatively, the temperature may be taken by placing a thermometer under the armpit.

Within the range
The value that is considered normal for body temperature is usually a mouth temperature of 37 degrees C. This is an average of values taken from a large number of people. A healthy person's body temperature might vary by as much as 1.5 degrees C above or below this value. Even in the same person, body temperature varies throughout the day. It is highest in the evening and lowest in the early morning.

In women there is another factor related to the menstrual cycle (see p. 224). At the time of egg release, there is a sharp increase in body temperature. Other causes of increase in body temperature are eating and exercise.

Sweat glands at work A body loses heat by radiation to cooler objects and gains heat from hotter ones. The amount of heat lost or gained will depend on the differences in surface temperature between objects and upon the nature of their surfaces. A shiny surface does not radiate as much heat as a dull black surface. The body behaves in a similar way. A black body is therefore a good radiator. This allows the body to lose or gain heat depending on the temperature of the surroundings.

The air in contact with the body will be warmed or cooled depending on whether the air or the body is warmer. **Convection** currents will be set up to aid the transfer of heat to and from the body. In windy conditions the rate of heat exchange by convection is increased considerably.

Heat can also be gained or lost by **conduction** through direct contact between the body and its surroundings. This form of heat exchange is not normally very important since air is not a very good conductor of heat. Conductive heat loss or gain is important when the body is in water because water is about 25 times better at conducting heat than air. This explains why you can stand undressed in a room at 21 degrees C and not feel cold. Yet you feel very cold immediately on diving into water at the same temperature.

The loss of heat from the body depends very much on the temperature of the skin. The body can control its heat exchange by regulating the flow of blood to the surface layers. This takes place by **dilating** (making wider) or **constricting** (making narrower) the large network of capillaries in the skin. There are also vessels which by-pass the skin and 'shunt' blood away from it when it is cold.

Heat that is generated in the body comes from chemical reactions and helps to keep the body warm. Heat is lost by **evaporation** of water from the lungs and from the skin. This accounts for about 25 per cent of the heat lost.

The sweat glands are capable of pouring large amounts of fluid over the body surface. This in itself does not get rid of heat. In order for this to happen the water must evaporate. If the air is very dry, water evaporation will take place rapidly. In humid conditions the sweat will stay on the body surface. To work in hot, humid conditions is therefore more stressful than working in a dry atmosphere, even though the temperature of the surroundings may be much higher.

The effects of cold can be equally unpleasant. Man has learned to adapt to a cold climate by means of his behaviour and technology. However, the form of the body has evolved under the selective pressures of the environment. This has given rise to the short stocky appearance of people exposed to cold climates and the tall thin shape of those living in hot climates (see Fig. 3.6.11).

Fig. 3.6.11 A Negro is tall and thin, an Eskimo shorter and stocky

Questions: Domains I and II

1 The diagram shows the structure of a kidney nephron.

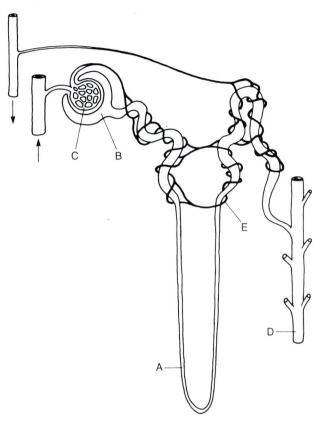

a Name the parts labelled A–E. (I)
b Where does filtration occur? (I)
c Name three substances that pass from the blood into the tubule. (I)
d Name one substance that does not pass from the blood into the tubule. (I)
e Name one substance which is completely reabsorbed into the blood from the tubule. (I)
f Name one substance which is not reabsorbed into the blood. (I)
g Where does the tube labelled D lead to? (I)
h What is the name of the liquid that passes out of the kidney? (I)
i Name two substances, other than water, which normally are present in the liquid leaving the kidney. (I)

2 The table shows average values for some of the substances filtered and reabsorbed each day by the kidneys.

Substance	Amount filtered	Amount excreted	Amount reabsorbed
Water	180dm^3	_____	178.2dm^3
Sodium	600g	_____	587g
Glucose	180g	_____	180g
Calcium	5g	_____	4.8g
Urea	50g	_____	20g
Potassium	35g	_____	33g

a Copy the table and complete it after calculating the amount of each substance excreted per day. (II)
b Name the region of a nephron (tubule) where most reabsorption of solutes occurs. (I)
c Glucose is actively reabsorbed but the reabsorption of urea is passive. Explain this. (I)
d Name the process by which water is reabsorbed. (I)
e Name the hormone which affects the amount of water reabsorbed. (I)
f Name the gland which secretes this hormone. (I)
g Which of the above substances would be excreted in larger amounts under the following conditions?
   i High protein diet.
   ii Lack of insulin.
   iii Removal of the parathyroid glands. (I)

3 At 11 am two people of similar age and size were given 1dm^3 of liquid to drink. Person W was given water but person S was given 1% salt solution, which is the same strength as body fluid. The table shows the urine volumes produced shortly after drinking.

Time	W	S
11.00	1dm^3 water drunk	1dm^3 salt solution drunk
11.25	380cm^3 urine	none
11.45	420cm^3 urine	none
12.30	200cm^3 urine	150cm^3 urine
14.00	120cm^3 urine	100cm^3 urine
Total	1120cm^3 urine	250cm^3 urine

a What is the response of the kidneys of W to drinking 1dm^3 of water? (II)
b Why was the total volume of urine passed by S so much smaller than the total volume of W? (I)
c State one other difference shown in the table, between the urine production of W and S. (II)
d Give two differences you would expect between the contents of a sample of W's and S's urine. (I)

4 An artificial kidney consists of blood flowing directly from a patient's body through cellophane tubing, immersed in a fluid, as illustrated in the simple diagram.

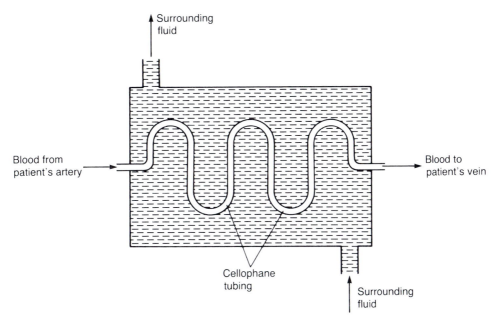

Surrounding fluid

Blood from patient's artery

Blood to patient's vein

Cellophane tubing

Surrounding fluid

a For what structure does the cellophane tubing substitute? (I)

b Explain how a salt e.g. sodium chloride, is removed from the blood using an artificial kidney. (I)

c A person undergoing dialysis must avoid certain foods. Name three substances which should be restricted, giving a reason in each case for your choice. (I)

Experimental skills and investigations: Domain III

1 What does the kidney filter out of the blood?

Procedure

1 Label three test tubes A, B and C. Place 2cm^3 of plasma, from centrifuged blood (from an abatoir), in tube A. Place 2cm^3 of urine in test tube B and place 2cm^3 of distilled water in test tube C.

2 Crush a tablet of urease for each test tube and add the urease powder to each tube. Place a stopper in each tube.

3 Place a piece of moist litmus paper in each tube, held between the stopper and the glass. Do not allow the litmus to touch the liquid.

4 Incubate the tubes at 37°C for one hour.

5 Record the colour of the litmus paper.

6 Test samples of A, B, and C with Clinistix and Albustix. Clinistix turns yellow in the presence of glucose. Albustix turns red if protein is present. Benedict's and Biuret tests can be used for reducing sugar and protein respectively, if Clinistix and Albustix are unobtainable.

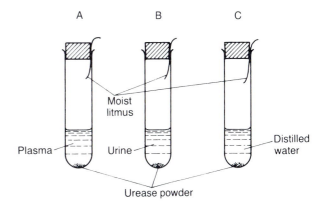

A   B   C

Moist litmus

Plasma   Urine   Distilled water

Urease powder

QUESTIONS

a List the substances proved to be present in A, B and C.

b Suggest the name of the substance that turns litmus the observed colour.

c Assume that blood plasma is filtered by the kidneys. What substances are in blood plasma that are not in urine?

3.7 Our 1000 K computer

Fig. 3.7.1 Our 1000K computer

Nervous control

It is the special development of the **nervous system** that makes us so different from the other animals. The nervous system carries messages that enable our thoughts to bring about actions. It also enables the automatic functions of the body to take place. The most important differences between our nervous system and that of other animals are to be found in the **brain**. This, plus the **spinal cord**, makes up the **central nervous system**.

It is made of special cells called **neurons**. Most of them are packed together in the brain and spinal cord, protected in the skull and the vertebrae. The central nervous system is linked with other parts of the body by the **peripheral nervous system**. It is made of bundles of long thin nerve fibres. Each is a thin part of a single neuron. The individual neurons behave in much the same way in all animals. Their function is to carry messages from one place to another. A process involving electrical changes takes place. Each nerve fibre conducts messages in one direction only. The nerve fibres in the peripheral nerves are (1) **sensory**, carrying messages to the central nervous system and (2) **motor**, carrying messages from the central nervous system. The basic plan is shown in Fig.3.7.2.

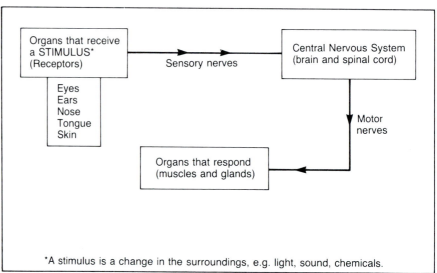

Fig. 3.7.2 The basic plan of the nervous system

Thoughtless actions – reflexes

The automatic action of a muscle or a gland resulting from stimulation of a **sense organ** (**receptor**) is called a **reflex**. Examples are the rapid withdrawal of the hand from something that causes pain, or the narrowing of the pupil when a light shines into your eye.

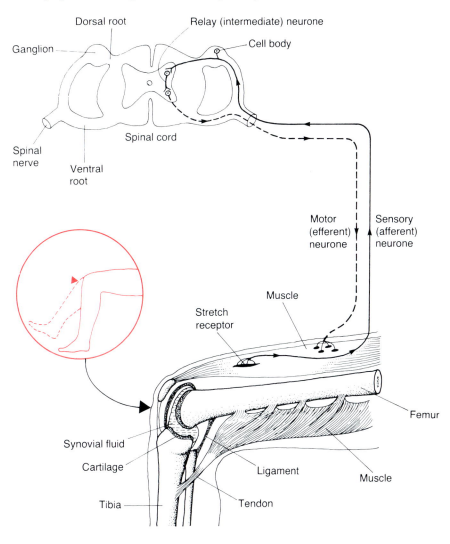

Fig. 3.7.3 The knee-jerk reflex

The **reflex arc** is like an electric circuit leading from a sense organ, to the spinal cord and back to a muscle or gland (see Fig.3.7.3). When the tendon of the knee is tapped, sensory receptors pass the impulse along a sensory nerve running to the spinal cord. Within the spinal cord, the impulse is passed on to another nerve cell whose fibre passes out of the cord to a muscle.

In this case the muscle will contract causing the lower leg to jerk upward. This is the most simple type of reflex arc, involing only two nerve cells. Sometimes there are three. One acts as a **relay nerve cell** between the sensory and the motor nerve cells. The junction between the fibres of one nerve cell and the fibres of the next is called a **synapse**.

Reflex arcs are not completely closed circuits. There are also nerve fibres to and from the brain which connect in the spinal cord, through synapses. These connections between the brain and the spinal reflex arcs make possible the most complicated muscular actions. The brain also receives impulses from the ears and eyes and combines these in reflex actions. Coordinated movements such as running and dancing then become possible.

Bridges between nerve cells

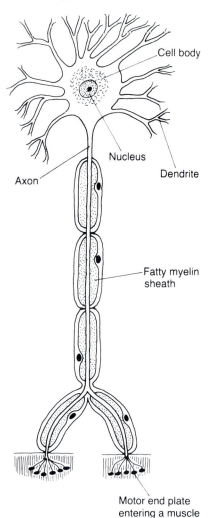

Fig. 3.7.4   A single nerve cell

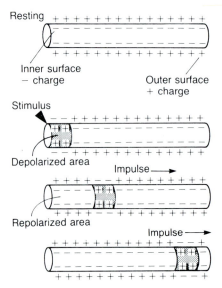

Fig. 3.7.6   How impulses travel

Nerve cells have fibres projecting from them. Branching fibres (**dendrites**) pass impulses to the nerve cell, and a long fibre (**axon**) transmits the impulse away from one nerve cell to the next (see Figs. 3.7.4 and 3.7.5). The fibres are insulated with a fatty myelin sheath to prevent electrical energy loss and are grouped into bundles. A synapse is the connection between the axon of one nerve cell and the dendrites of the next. At the synapse, the impulse can either be transmitted or blocked.

A synapse is really a minute gap into which chemicals are secreted by the tip of the nerve cell. The type of chemical secreted either allows the impulse to pass across or it stops it. The synapse therefore acts like a bridge about one-millionth of an inch long.

In a simple reflex arc the impulse is passed across the synapse, goes on to the next synapse, and so on around the circiut. Eventually it activates the muscle or gland. All synapses are permanently either open or closed. This simple 'yes' or 'no' system of passage allows the contraction of one group of muscles while another antagonistic group (see p. 96) relaxes.

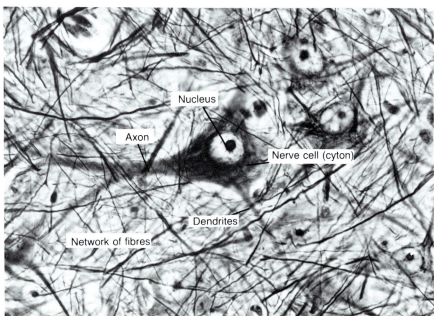

Fig. 3.7.5   Photomicrograph of nerve cells

How the messages are carried
The message is carried by the axon, a long tube-like part of the nerve cell. The liquid inside this tube is separated from the surrounding liquid by a membrane. The chemicals in the liquids inside and outside differ. There is a voltage difference between the inside and outside of about one-tenth of a volt. Electric currents can alter this voltage and thus change the membrane (see Fig.3.7.6). If the change is large enough there is a sudden alteration in the voltage across the affected region of the membrane. This change is called the **action potential**.

Nerve fibres are very narrow compared to their length. A large fibre may be less than twenty microns in diameter and over one metre long. One part of an axon may be active while other parts are resting. The voltage across the cell membrane can be different at different places along the axon. A current will then flow. This will produce activity in another part of the axon that was previously resting. The activity is thus carried from one end of a nerve fibre to another.

The brain – centre of our universe

The brain is as firm as a thick blancmange, contained within the skull (see Fig. 3.7.7). At the base of the skull is hole through which the spinal cord passes. Both the brain and spinal cord are protected by three covering layers called the **meninges**. The outermost of these layers is the thickest. It lines the inside of the skull, is very fibrous, and is called the **dura mater**. The middle layer and the innermost layer are close to both brain and spinal cord, and are thin and delicate. They are called the **arachnoid mater** and the **pia mater**. Between the two there is a space that contains fluid – the **cerebrospinal fluid**. This fluid bathes the surface of the spinal cord and the brain. It acts like lymph taking material to and from the nerve cells. See Fig. 3.7.8.

Fig. 3.7.7 Dorsal, ventral and half-section views of the brain

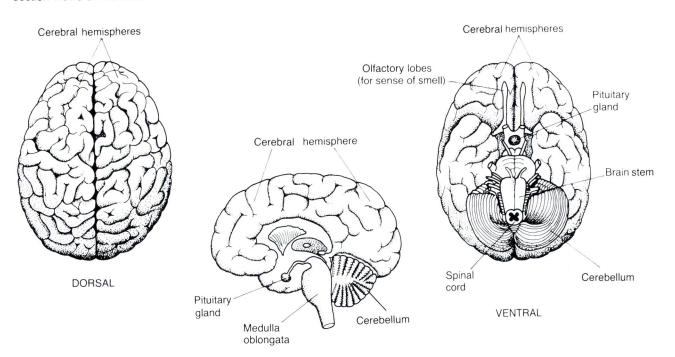

Cerebral hemispheres

DORSAL

Cerebral hemisphere

Pituitary gland

Medulla oblongata

Cerebellum

LATERAL

Cerebral hemispheres

Olfactory lobes (for sense of smell)

Pituitary gland

Brain stem

Spinal cord

Cerebellum

VENTRAL

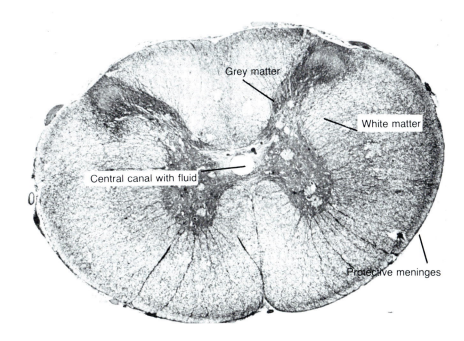

Grey matter

White matter

Central canal with fluid

Protective meninges

Fig. 3.7.8 Photomicrograph of transverse section through the spinal cord

The brain consists of the **cerebral hemispheres** and **midbrain**. Behind these are the **cerebellum** and the **brain stem**. The brain stem consists of the **pons** and the **medulla oblongata**. It joins the brain to the spinal cord.

The cerebral hemispheres make up the largest part of the brain. Each hemisphere is divided into four lobes. They contain the centres from which impulses are sent to muscles and glands. They also receive the sensations of heat, cold, pain and balance from the sensory nerves.

The brain appears greyish at the outer edges (the **cortex**) and white inside. The grey colour is due to the massive collection of brain cells. The white areas contain the nerve fibres which are insulated with white fatty material. These nerve fibres connect nerve cells with one another in various parts of the brain. The nerve cells are like the transistors of a computer. The nerve fibres represent the wires connecting all the transistors. Some of the fibres are very long. They run from the brain all the way down the spinal cord carrying impulses as far as the legs.

Unconscious control – the autonomic nervous system

The system of nerves spreading through the body is called the **peripheral nervous system**. The outer or **somatic** part links up the sense organs that provide us with a 'picture' of the outside world, with the effectors that respond to our advantage. In order that these sense organs and effectors function properly, the inner machinery of the body must function smoothly. The rate at which it works must be adjustable so that the varying demands of the outer structures may be satisfied. When we move our muscles need a greater supply of oxygen than when we are resting. The heart must beat faster to supply more blood. Food and oxygen must reach the muscles in great quantites, and the lungs must be filled and emptied more rapidly than usual. The blood vessels to the muscles become wider to supply the extra blood.

Nerves that coordinate all of these internal changes form the **visceral** part of the peripheral nervous system, the **autonomic system**. The system can be divided into two parts, **sympathetic** and **parasympathetic**. Generally their actions are opposite. Most organs receive both sympathetic and parasympathetic nerves, though some, for example sweat glands, receive only one – the sympathetic. The alimentary canal, apart from the sphincter muscles, relaxes when it receives signals through sympathetic nerves. Signals through parasympathetic nerves cause it to contract. The sphincter muscles are affected in the opposite way, contracting on the receipt of impulses from the sympathetic nerves and relaxing on receipt of signals from the parasympathetic nerves.

Signals from the sympathetic nerves increase the strength of heart beat but the muscles of the main arteries leaving the heart relax, so allowing an increase in blood flow. Sympathetic nerves usually cause the muscles of the rest of the blood system to contract. During digestion the blood flow to the gut is increased by the gut blood vessels relaxing. At the same time the blood supply to the rest of the muscles of the body is reduced.

Some of the actions of the autonomic nervous system are summarized in the table opposite. The sympathetic part is generally concerned with preparing us for violent action and for abnormal conditions (see Fig.3.7.9). The parasympathetic system is more concerned with re-establishing normal conditions after we have been involved in strenuous actions.

Some of the actions of autonomic nerves

Organ	Sympathetic causes	Parasympathetic causes
Heart	increase in rate and strength of beat	reduction in rate and strength of beat
Skin blood vessels	constriction	no supply
Blood vessels to muscles of body	usually constriction	no supply
Gut blood vessels	constriction	no supply
Muscles of gut except sphincters	relaxation	contraction
Sphincters of gut	contraction	relaxation
Salivary glands	production of mucus	production of saliva
Pancreas	no supply	production of pancreatic juice
Hair muscles	contraction	no supply
Sweat glands	release of sweat	no supply

Fig. 3.7.9 The effects of the sympathetic nervous system

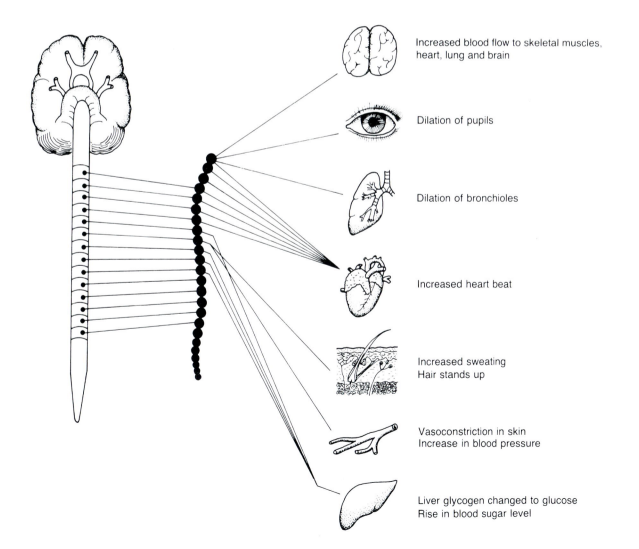

Increased blood flow to skeletal muscles, heart, lung and brain

Dilation of pupils

Dilation of bronchioles

Increased heart beat

Increased sweating
Hair stands up

Vasoconstriction in skin
Increase in blood pressure

Liver glycogen changed to glucose
Rise in blood sugar level

Chemical control through chemical messengers

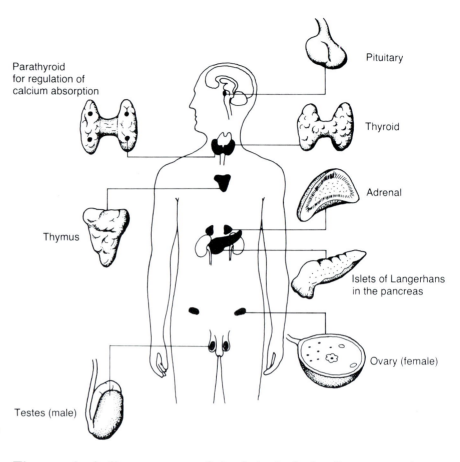

Parathyroid
for regulation of
calcium absorption

Pituitary

Thyroid

Adrenal

Thymus

Islets of Langerhans
in the pancreas

Ovary (female)

Testes (male)

Fig. 3.7.10   Location of the endocrine glands

The endocrine system

There are basically two groups of glands in the body. One group releases secretions through ducts and the secretions act near the site of release. These are the **exocrine glands**, for example sweat glands and salivary glands.

The other group, the **endocrine glands**, secrete chemicals directly into the blood stream (see Fig.3.7.10). These chemicals are called **hormones** and they circulate in the blood. They may affect parts of the body that are far removed from the glands that produce them. Hormones can be described as **chemical messengers** which regulate almost all of the chemical actions taking place in the body.

Control of blood sugar

One very important function of the endocrine system is the control of the concentration of glucose in the blood. It is essential that the concentration of glucose in the blood remains at about 0.1g per 100cm^3. If it falls much below this, the brain is unable to have enough glucose for energy release. Glucose levels above 0.1 per cent by weight (i.e. 0.1g per 100cm^3) also disturb the body functions. In particular, the ability of the kidneys to reabsorb glucose is lost and glucose appears in the urine. The body is eventually drained of fuel.

Small groups of cells in the pancreas called the **islets of Langerhans** secrete a hormone called **insulin** into the circulating blood. One function of insulin is to help the transport of glucose from the blood stream into cells. Insulin also helps the formation of glycogen from excess glucose in the liver. Glycogen is the animal equivalent of starch and, because it is insoluble, can be stored in the liver cells as a reserve of carbohydrate. When it is needed it is changed back to glucose. It can be stored in the liver in limited amounts. If it exceeds the limit, it is changed to fat, which can be stored in almost unlimited amounts. Insulin speeds up the manufacture of fats.

The secretion of insulin from the pancreas is controlled by the blood glucose concentration. When blood glucose rises after a meal, the pancreas senses this increase. It then increases its rate of insulin secretion. Insulin then causes excess glucose to be taken out of the blood circulation. It does this by stimulating its use by cells in respiration or by changing it to glycogen to be stored in the liver and muscles. The concentration of glucose in the blood is therefore reduced to its normal level.

If blood glucose falls, for example after exercise, the pancreas again senses the change and reduces the secretion of insulin. It also secretes another hormone called **glucagon**. Its effect is to break down glycogen in the liver to form glucose. The combined effect of reduced insulin and increased glucagon is a rise in blood glucose.

Sometimes the islets of Langerhans fail to produce insulin or the body cells fail to react to insulin. This causes a condition called **diabetes mellitus** or **sugar diabetes**. It can occur because of several reasons. In young people insulin may not be produced or the body may not react to insulin in the normal way. Middle-aged overweight people sometimes produce too little insulin to meet the demands of the body. Under conditions of stress, the body may fail to respond to insulin at its normal levels in the blood. People suffering from diabetes are treated by regular injections of insulin obtained from other animals. Carbohydrate intake must be reduced to lessen the concentration of blood glucose. Emergency treatment for a person who has fainted because of having too little blood glucose is to administer an immediate glucose supply.

Fight or flight – the adrenal glands

The adrenal glands each consist of two parts, the **cortex** and the **medulla**. The medulla secretes a hormone called **adrenalin**. A certain amount of adrenalin is secreted into the bloodstream all the time. However, its effects are most noticeable under conditions of fright or stress. For this reason, it is often called the 'fight or flight' hormone. Its function is to help you to face a stressful situation. It raises the blood pressure, increases the rate of heart beat and its stroke volume, increases the rate and depth of respiration, and raises the blood sugar concentration.

All of these responses are directed towards giving you energy to react to a harmful situation. The increased blood flow will bring more fuel and oxygen to muscle cells so that energy can be released more readily. The energy is then used for escaping or for withstanding the stressful problem.

The cortex of the adrenal glands produces several hormones. One of these is **aldosterone** which helps to regulate sodium concentration. Another is **cortisone** which helps the body to adapt to stress.

Regulation of growth – the thyroid

There are some people who find it almost impossibe to lose weight despite controlled intake of food (dieting). Others find it difficult to put on weight even if they eat a lot. It is the **thyroid gland** that may cause these problems. It secretes a hormone called **thyroxine** which speeds up the rate at which glucose is used to release energy. If the thyroid gland is underactive, it will not produce enough thyroxine to cause normal breakdown of glucose. If this happens, the glucose can be changed to fat and the person will become overweight. The opposite is the case of people with overactive thyroid glands.

People with severely underactive thyroid glands become very stunted in growth and also mentally retarded. This results in a condition known as cretinism. They are often given regular doses of thyroxine to overcome the problem.

The master – the pituitary gland

The **pituitary gland**, at the base of the middle of the brain, controls the secretions of many other endocrine glands (see Fig.3.7.11). Because of this control, it is often called the 'master' hormone gland. Some of the hormones it secretes have a direct effect on cells. For example, the growth hormone causes a direct increase in growth rate. The effects of the pituitary gland are shown in Fig.3.7.12.

Other endocrine glands include the **gonads**. The effects of these are described in the section on reproduction (p. 224).

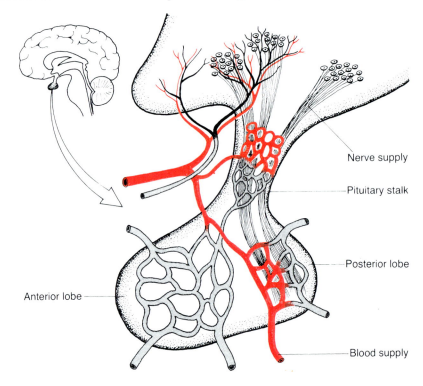

Fig. 3.7.11   The pituitary gland

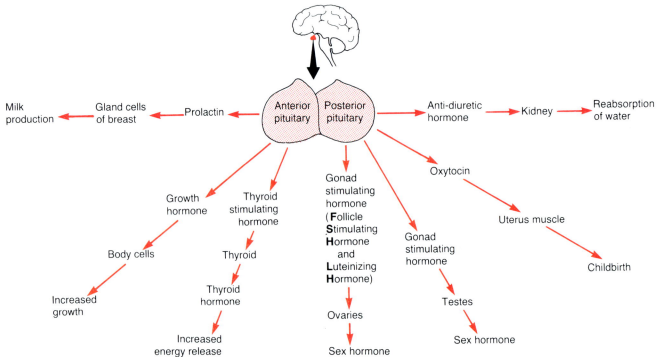

Fig. 3.7.12   The effects of the hormones of the pituitary

**Our own detective agency –
the senses**

*Our window on the world –
the eye*

The eye is the organ of sight. It is located in a bony socket, the **orbit**, which protects it from injury (see Fig.3.7.13). Further protection is provided by fat in which it is embedded. In humans both eyes have an overlapping field of view for three-dimensional (stereoscopic) vision.

Each eyeball is a near-perfect sphere, about 24mm in diameter. It keeps this shape as a result of tough surrounding layers confining fluids within. Each eye has three coats: (1) a fibrous **sclera** which has a transparent front, the **cornea**; (2) a vascular coat, consisting mainly of blood vessels and made of the **choroid, ciliary body**, and **iris**; and (3) a layer of sensitive nervous tissue, the **retina**.

The sclera is a tough membrane mainly responsible for keeping the shape of the eye. Its surface is glistening white. Through the cornea, the iris can be seen. The cornea bends light rays so that an image is focused on the retina. In this respect it is very much like a camera. Tears bathing the covering (**conjunctiva**) have an antibacterial effect and are produced by tear (**lachrymal**) glands near the eye.

The choroid consists of a network of small capillaries. It is responsible for the nourishment of the internal structures of the eye, especially the retina.

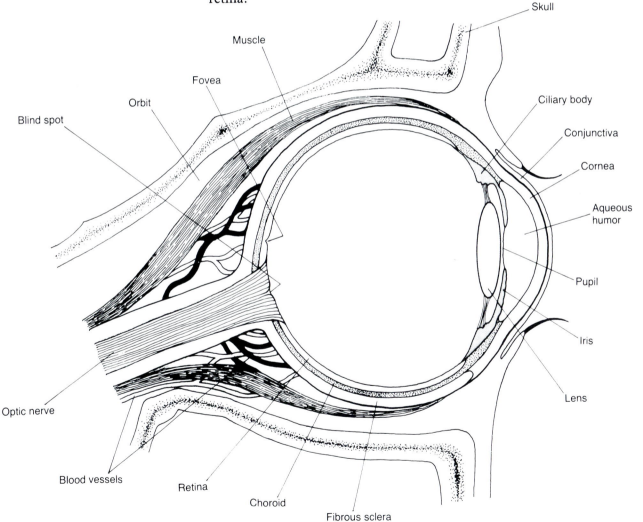

Fig. 3.7.13   A section through the eye in its orbit

Regulating the light

The ciliary body is the thickened front of the choroid. It runs like a ring around the inside of the eye and connects the choroid with the iris. In the ciliary body are the **ciliary muscles** and many blood vessels. The muscles alter the tension on the lens using the **suspensory ligaments**. These anchor the **lens** in place behind the iris. The ligaments are attached to the **lens capsule** and to the ciliary body.

The iris is circular with a central hole, the **pupil**. The size of the pupil is regulated by the amount of light entering the eye. As with the camera, a smaller hole for light gives a greater depth of focus than a large one.

Alterations in the size of the pupil are made by circular and radial muscles. They are controlled by the autonomic part of the central nervous system (see p. 186). The sympathetic nerves cause enlargement, and the parasympathetic fibres produce constriction. The change is a reflex action. Bright light results in constriction and fear causes dilation (enlargement).

The iris divides the part in front of the lens into two chambers. Both chambers are filled with a clear fluid called the **aqueous humor**, due to its watery nature. The colour of the iris varies between individuals and in some cases between the eyes of the same individual. The colour (pigment) is an additional protection against bright light damaging the retina.

The sensitive film

The retina can be compared with a very sensitive film in a camera. All the other structures inside and outside the eyeball focus light rays on it, or nourish and protect it. It receives reversed and inverted images. These are sent to the brain as impulses along the **optic nerve**.

The eye can also be compared with a television aerial. If the signal (the image on the retina) is strong and clear, then vision (the picture on the television set) will be good. The retina is a delicate membrane about 0.4mm thick, with its outer surface touching the choroid. Its inner surface touches the jelly-like substances filling the part behind the lens. This gel is called the **vitreous humor**. The retina connects with the optic nerve at the back of the eye.

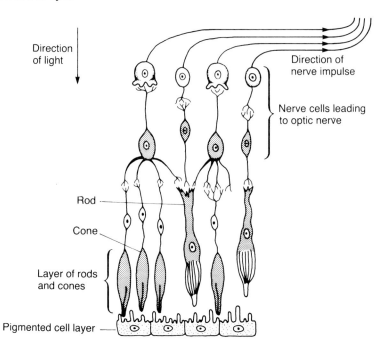

Fig. 3.7.14   Structure of the retina

Although the retina is only 0.4mm thick it has separate layers. The most important is the layer of nerve cells called **rods** and **cones** (see Fig.3.7.14). Each rod and cone is sensitive to light. Cones are responsible for colour vision. They only work in bright light. Rods work in dim light.

The retina produces a purple chemical called **visual purple**. It is bleached when exposed to light and is like the chemical emulsion of a film. Near the centre of the retina is an oval, yellowish area, 1.5mm in diameter. This is the most sensitive area concerned with detail, and sharpness of the visual image.

There is a small central depression in the retina called the **fovea centralis**, which is the most sensitive part of all. The region where the optic nerve leaves the eye consists of nerve fibres with no cells. It is not sensitive to light and is called the **blind spot**.

Focusing the rays
Very few people have perfect eyesight and most will need spectacles at some time in their lives. An eye with no defects is able to focus an image both at close range and at a distance, without any strain. A printed page will look as clear and sharp as a distant star, because of the way the lens can focus. The purpose of the lens is to bend light rays on to the retina so that they arrive there as a sharp image. The lens can alter its shape so that the images of both near and distant objects can be focused.

The lens is made of fibres inside an elastic capsule. The fibres are arranged in such a way that when the eye is doing close work, the lens is fat and when the eye is relaxed, looking into the distance, the lens is thinner. In order to produce an image of a distant object on the retina, the ciliary muscles relax. The suspensory ligaments become stretched, making the lens flatter. This allows parallel rays of light entering the eye to be focused. When focusing a near object, the ciliary muscles contract, relaxing tension on the suspensory ligaments. The elastic lens then becomes almost spherical. This process is called **accommodation** and is automatic (see Fig.3.7.15).

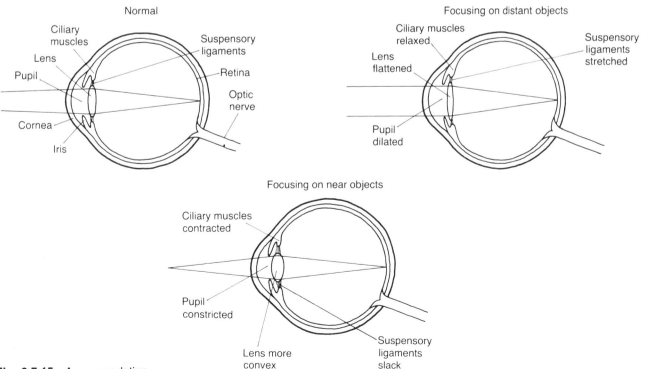

Fig. 3.7.15   Accommodation

Short and long sightedness

One cause of short sightedness may be the direct result of modern living – watching television, reading and so on. Short sightedness can also be due to a defect of the retina, or more commonly, due to the eyeball being too long from front to back. The difficulty comes if distance viewing is required. The lens cannot accommodate itself by making itself thin enough to focus the distant object without strain, and spectacles are needed.

Without spectacles the eye would try to focus the distant object as though it were much nearer and the result would be a blurred image on the retina. A **concave** lens of the correct shape will spread out the light rays making them seem to come from a nearer object, so that the eye can obtain a clear image (see Fig.3.7.16).

Long sight can be caused by the eyeball being too short from front to back or by the hardening of the lens. A long-sighted person can focus distant objects but not close ones (see Fig.3.7.16). **Convex** (converging) lenses are worn to bring together the light rays so that a clear image can be obtained on the retina.

As you grow older the lens hardens. The ciliary muscles do not have the power to make the lens spherical enough to focus near objects. Special reading spectacles are needed, similar to those needed by long-sighted people.

Fig. 3.7.16   Eye defects

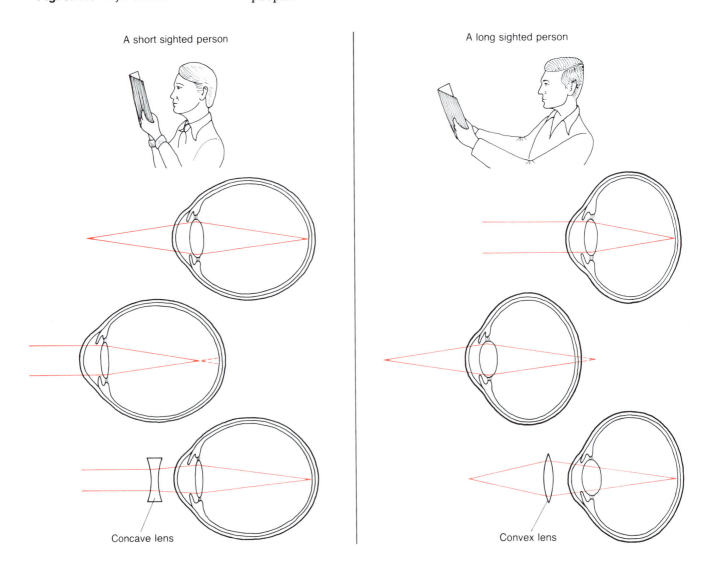

A short sighted person

A long sighted person

Concave lens

Convex lens

The ear   The human ear is a delicate organ. Parts control the sense of balance and other parts receive sound signals. These are passed as impulses along a nerve to the brain so that we can hear. When a stretched string is plucked it will vibrate and make a sound. The vibrating string pushes and pulls the air around it. In this way vibrations are transmitted through the air to our ears.

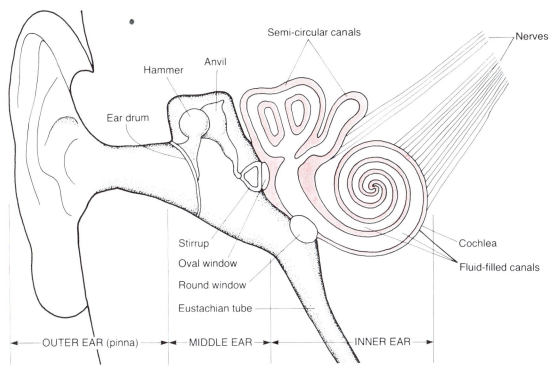

Fig. 3.7.17   A simplified diagram of the ear

Sound waves are collected by the **outer ear** called the **pinna** (see Fig.3.7.17). The waves make the **ear drum** vibrate. This movement of the ear drum is carried through the air-filled **middle ear**. Three tiny bones, the anvil, hammer and stirrup, pass the vibrations to the **oval window** at the innermost side of the middle ear. Beyond the oval window is the **inner ear**. The part of the inner ear concerned with hearing is called the **cochlea** (see Figs. 3.7.18 and 3.7.19). It is a coiled tube filled with fluid. Vibrations of the oval window produce pressure changes in the fluid in the cochlea. This affects tiny, sensitive hairs that link up with nerve fibres in the Organ of Corti (see Fig. 3.7.20). The movement of these hairs results in impulses being sent along the nerve fibres to the **auditory nerve** which carries the impulses to the brain. In the brain these impulses are translated as sound. The ear does not hear. It just receives sound waves which are transmitted as impulses to the brain. It is the brain that interprets these as sound.

Fig. 3.7.18   The cochlea straightened out

COCHLEA SPIRALLED

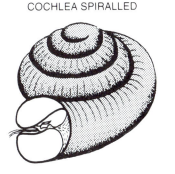

COCHLEA UNCOILED

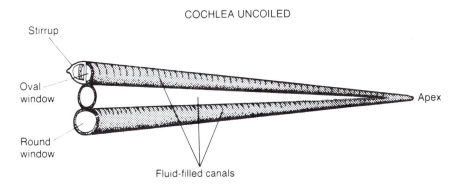

Fig. 3.7.19   Section through the cochlea

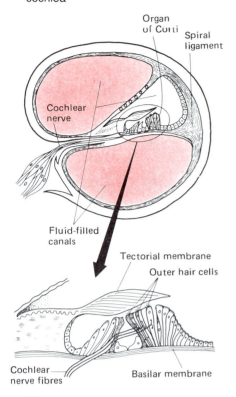

Fig. 3.7.20   The organ of Corti

The middle ear is connected to the back of the throat by a narrow tube – the **eustachian tube**. This is normally closed by a small muscle. When we swallow or cough the tube opens to let air into the middle ear from the throat. This ensures that the pressure in the middle ear is the same as that in the outer ear, so that equal pressures are maintained on both sides of the ear drum.

The area of the ear drum is nearly thirty times that of the oval window. The bones of the ear form a system of levers which 'gear down' the movements of the ear drum. This size difference between the two membranes, and the gearing down, result in the pressure on the oval window being twenty-two times that of the original pressure on the ear drum.

The ear is protected from very loud sounds by the action of two muscles. One is attached to the ear drum. The other is attached to the **stirrup**. When these shorten, the ear drum and the oval window (to which the stirrup is attached) become stretched so that their to-and-fro movement is reduced.

The wall of the cochlea is bony. Since the cochlea is filled with fluid and fluids cannot be compressed, the to-and-fro movements of the oval window cause pressure changes in the fluid. These pressure changes cause the sensitive hair cells to vibrate. The stimulation of the hairs on the hair cells causes impulses to pass along the ear nerve to the brain.

If the cochlea is straightened out, it can be seen to be a tapered tube containing three fluid-filled canals, the semi-circular canals (see Fig.3.7.18). The two outer canals are in communication with each other through a small opening at the end of the cochlea. The oval window closes the end of one of these canals. Another membrane, the **round window**, closes the other. When the oval window bulges inwards (i.e. is pushed by the stirrup) the pressure change in the fluid in the canals causes the round window to bulge outwards. This is safety device which provides pressure relief.

Taste and smell   The body has a system of receptors that give it information about chemicals. It consists of the organs of taste and smell. **Taste receptors** occur mainly on the tongue. A few are elsewhere in the mouth, pharynx and on the epiglottis. Small projections on the tongue have flask-shaped **taste buds** scattered over them. These consist of groups of cells that have nerve endings between them and wrapped around them. When the taste buds are stimulated, signals pass along the nerve fibres to the brain. The taste buds are moistened by **saliva**. Chemicals must be in solution before they can stimulate the receptor. During dry cold weather the senses of smell and taste may be very much reduced.

For each of the four types of taste – salt, sweet, acid (sour), and bitter – there are taste buds in different parts of the tongue. The tip is most sensitive to sweet and salty substances; the sides to acid substances; while the back of the tongue is most sensitive to bitter substances.

The texture of food and its temperature will also affect the taste of food. Touch and temperature receptors in the mouth will be stimulated.

We can of course appreciate tastes other than the four basic ones. These are part of the sense of smell, dependent on receptors in the back of the nose. The sensitive cells are embedded in the lining epithelium of the nose. They are in the **nasal cavity**. The smell (**olfactory**) cells have axons passing to the central nervous system (see Figs. 3.7.21 and 3.7.22).

Generally, nerve processes grow out from the central nervous system. They are surrounded by supporting cells and glands which produce a fluid. This moistens the ends of the olfactory cells. The chemicals to which the receptors are sensitive dissolve in this fluid. All the olfactory cells have the same structure yet we can appreciate a wide range of smells. There are relatively few taste receptors but there are many smell receptors sensitive to different smells.

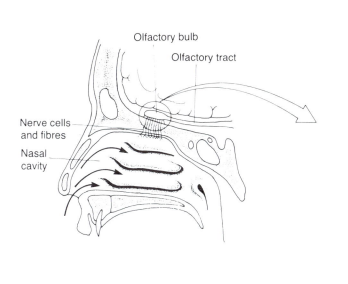

Fig. 3.7.21 The nose

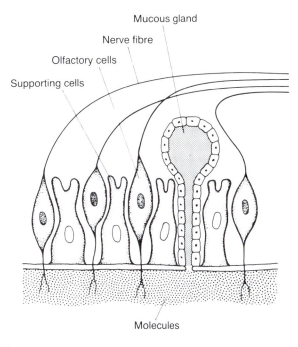

Fig. 3.7.22 Close up of the organ of the sense of smell – the olfactory bulb

Protecting the mind It is becoming clear that many mental disorders can be prevented. There are opportunites to prevent related problems such as the misuse of alcohol, drug abuse and attempted suicide. In addition, many mental diseases could be significantly influenced by changes in the behaviour of people.

In a population of 100 000 people in a developing country (and the figures are similar in developed countries), there are at least 6000 to 8000 people who have mental disorders or who depend on alcohol or drugs. Many countries have been taking serious steps to do something to prevent alcohol problems. The action had changed from treating alcoholism to early detection of problem drinking. Many people who would refuse to accept being called alcoholics will much more readily agree that they have problems with the way they drink.

Prevention of alcoholism can be achieved by self-help groups such as Alcoholics Anonymous, in some countries, or by other community-based organizations. They often concentrate on youth because this is the stage where normal or social drinking can easily change into troublesome or heavy drinking.

Adolescence seems to be the best stage to focus on because many young people experiment with alcohol. They often want to emphasize their personal identity in defiance of their parents, or as a password to relate better with older people, or as a way to relax in front of the opposite sex. It oftens results in many cases where younsters start to drink in excess. They can end up being cared for by their parents and being isolated from other people.

Prevention of drug problems

Throughout the history of mankind, there is hardly any country or culture which has not encountered problems with the use and abuse of drugs. In our time, drug problems have become varied and complex. The use of natural drugs such as cannabis, cocaine, khat and opium, has now reached epidemic levels.

Manufactured drugs such as amphetamines, barbiturates, sedatives and tranquillizers have become more readily available. Added to these are the growing habits among children and adolescents of sniffing solvents and inhaling paints and glues. There has also been a general rise in the use of drugs in most countries. Research shows that more people are now combining drugs and alcohol.

Cocaine misuse needs special attention. It is the most available drug of its kind. Its misuse is reaching an epidemic level in some regions of the world. Opium-eating among rural South-Asian populations and in certain Middle East countries has developed into the much more dangerous use of heroin in the form of smoking or by injection.

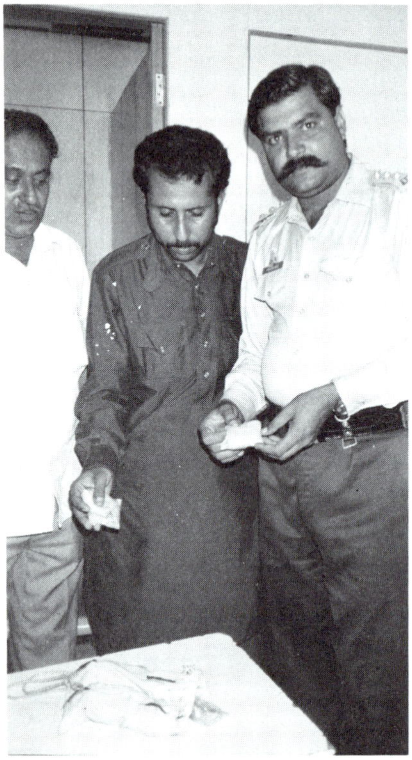

The drug scene is accompanied by a wide range of problems including criminality, and violence. It also is one of the most serious social health hazards that developed nations have ever had to face.

What should be prevented and who should do the preventing? In practice, efforts are generally directed towards controlling the supply of drugs. There are the international drug-control treaties to control drugs like cannabis, cocaine, opium, amphetamines, barbiturates and mescalin. At national level, there are the enforcement actions which may be undertaken by the police and customs.

Also there are measures for controlling the supply by restricting the availability of prescribed drugs. Public information and health education are approaches aimed at reducing drug demand (see Fig.3.7.23). Treatment for those who are dependent on drugs is now being suggested as a means of prevention. On the whole, the best preventive measures are those which are developed with due regard to culture and social life. A religious group may be protected against the abuse of alcohol and tobacco because of its beliefs.

Many countries have been trying very hard to combat drug problems. The fight is based on changes in attitudes and values. Countries can also learn from successful national anti-smoking campaigns. These have had an effective impact on personal and cultural attitudes. Drug problems are among the most damaging menaces of modern life. Their effective prevention calls for countries to adopt an all-out effort, involving clear long-term policies which are linked to active community cooperation.

Fig. 3.7.23 Drug smuggler caught in the act

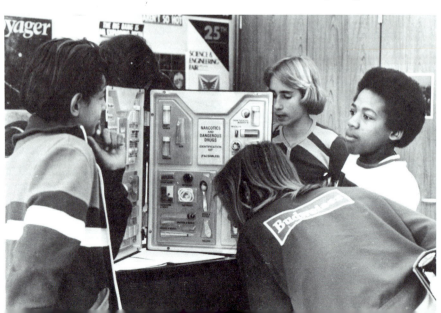

Fig. 3.7.24 Teenagers on drug information course

A summary of some of the major drugs used by people and their effects

Class	Name	Effects	Dangers
Sedatives	Barbiturates. Morphine	Relaxation and relief of anxiety.	Addiction. Drowsiness.
Stimulants	Caffeine. Drinamyl	Quickening of responses.	Addiction.
Narcotics	Cannabis	Affects the nervous system.	Addiction. Hallucinations. Lung cancer. Deformities in the foetus during pregnancy.
	Cocaine	Relief of pain.	Addiction. It is poisonous and causes paralysis.
	Opiates	Affect the nervous system.	Addiction. Hallucinations.
	Heroin	Affects the nervous system.	Chronic addiction. Mental and moral deterioration.
	LSD (lysergic acid diethylamide)	Affects the nervous system.	Addiction. Brain damage. Hallucinations. Foetal malformities.

Life's crises – stress   The effect of stress on the body can be to upset the proper functioning of many of its organ systems. In particular the heart, blood system, and the body's immune systems may be severely affected.

There are several ways to help reduce the chances that stress will lead to physical or mental disorders. However, many kinds of stress cannot be removed or escaped from. Yet we can still do something about them. For example:

1   Helping ourselves to adapt better to changed situations.
2   Changing the way we look at and evaluate situations.
3   Practising dealing with stressful situations.

The presence of friendly people around us to listen to our problems, give support, and actively help, can have a large influence on helping us to remain strong enough to resist illness. Another approach to stress reduction has been known and practised, particularly in Asia, for centuries. People who regularly practise meditation or relaxation techniques can produce a slower heart rate and a decline in blood pressure, as well as an increased feeling of calmness and well-being (see Fig. 3.7.25). Physical exercise, like a brisk walk or vigorous sport, is also a healthy way to relieve tension.

Fig. 3.7.25   Relaxation techniques can relieve stress

Questions: Domains I and II

1

a Name the structure that:
  i carries nerve impulses from the eye to the brain;
  ii is protected by the vertebral column;
  iii is made up of rods and cones. (I)

b Give two differences between a sensory neurone and a motor neurone. (I)

c Surgical operations can be performed to replace damaged ear drums with artificial ones. People who have just had this operation should not blow their noses violently. Explain why this is so. (I)

2

a What makes up the central nervous system? (I)

b During childbirth mothers are sometimes given an epidural. This is an anaesthetic injected into the lower part of the spinal cord. The effect is to prevent the mother feeling pain, although she continues muscular contractions of the uterus. Explain why the mother does not feel pain. (I)

3 Listed below are various stages, A to G, involved in a voluntary action such as picking up a magazine from a table. The stages are not in the correct order. Rearrange them so that they are in the correct order. (I)

A Impulses pass along the spinal cord.
B Light rays from the magazine enter the eye.
C Impulses pass along the spinal nerve to the muscles of the arm and hand to pick up the magazine.
D The vision area of the brain receives the image of the magazine.
E Impulses pass along the optic nerve to the brain.
F Impulses are sent from the motor area of the brain.
G Rods and cones are stimulated.

4

a In an experiment on the working of the nervous system a person had to press a button as soon as possible after a light flashed on. The time taken for the person to react to the light by pressing the button was recorded. The experiment was repeated 25 times. The graph shows the results.

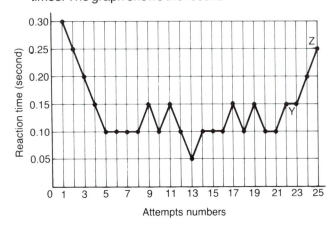

Attempts numbers

  i What was the fastest reaction time? (II)
  ii What was the slowest reaction time? (II)

b Describe what these results tell you about the reaction of the person over the course of the 25 tests. (II)

c Suggest an explanation for the shape of the graph between points Y and Z. (I)

d If the experiment had been carried out for another 10 attempts, describe what result you would expect, using the information on the graph to help you reach your conclusion. (II)

e Suggest what the effect of a large quantity of alcohol would be on the reaction time and explain your answer. (I)

5 Read the tables and answer the questions that follow.

Total number of known drug addicts

Age group	Males	Females	Total
Under 17	6	2	8
18–21	166	71	237
22–25	563	140	703
26–29	412	84	496
30–34	116	47	163
35–49	121	44	165
50+	80	128	208
Total	1464	516	1980

Number of drug addicts by type of drug

	1972	1973	1974
Methadone alone	624	731	1305
Methadone and heroin	250	241	239
Heroin alone	170	140	109
Other drugs	300	314	327
Total	1344	1426	1980

a Which age group has the greatest number of drug addicts? (II)

b Which drug has shown the greatest increase in drug addicts from 1973–74? (II)

c What is the medical use of a tranquillizer? (I)

d Alcohol and tobacco are two other drugs to which people become addicted. Where in the body does the absorption of alcohol take place? (I)

e Which three organs of the body may be damaged by excess alcohol? (I)

f Name two dangerous substances in cigarette smoke. (I)

g Which of these affects the cilia of the bronchi? Explain what it does to them. (I)

h Give one measure introduced by the Government to discourage people from smoking. (I)

6 Read the following passage carefully and then answer the questions about it:

Most drugs have an effect on the nervous system; some increase its activity (the stimulant drugs) while others (the depressant drugs or sedatives) reduce its activity.

Many drugs, such as alcohol and opium, have been used for many years to produce a happy state, and the dangers of using them were not realized. Other drugs such as pep-pills were first used medicinally and as such are of great value. When used indiscriminately they are very dangerous.

The chief danger of drugs is that they can be habit forming. A person may well try a drug for fun and become addicted to it remarkably quickly. This addiction begins simply as a desire to repeat the pleasant sensation, but it leads to a condition in which the body metabolism is altered so that it needs the drugs to function properly, as it did before the drug was taken. This is the true state of addiction.

The consuming of alcoholic drinks is recorded throughout history. Even today, in some countries they are the only safe drinks to take when the local water supply is unfit.

The intake of alcohol can have many different effects on the working of the body. Even a small amount affects the nervous system, making reactions slower with loss of judgement so making the possibility of accidents greater. Here is the reason for the use of the breathalyzer test for motorists suspected of drinking too much. The alcohol reaches the cerebellum of the brain with the result that muscular coordination and balance are affected. Alcohol circulating in the blood system causes a widening of blood vessels in the skin with the result that there is an increase in the loss of heat from the skin, although for a time the person does feel warmer because of the increased heat supply to the skin. The increased blood supply to the skin can cause less blood flowing to other organs so that the kidneys may cease to function properly. Heavy drinking over a long period can cause permanent liver damage.

Some drugs are called depressants because their general effect is to decrease the basic body metabolism, including slowing the heartbeat and rate of breathing. One commonly used type of depressant is aspirin and its related compounds which are very useful for the relief of pain and reducing temperature. If taken in excess or over a long period of time they can cause defects of the nervous system, sickness, irregular heartbeat and damage to the stomach lining. Although aspirin is readily available it should not be used for minor pains. Morphia and heroin are powerful pain killers but a person can easily become addicted to them, with the usual result of a complete breakdown in personality.

The stimulants cause an increase in body metabolism. One of the commonest stimulants is caffeine found in small amounts in coffee and tea, but a person would have to drink very large amounts of these for the drug to have any serious effect. Cannabis, if taken in large amounts, can cause confusion and other mental defects. Amphetamines are the base of many pep-pills and may be used to help a person over a difficult period of their life, but one can easily become addicted to them with the result of irresponsible behaviour.

a What does this text describe as being one of the chief dangers of the drugs? (II)

b Give three effects on the body of taking a depressant drug. (II)

c Name a commonly used depressant mentioned in the passage. (II)

d What are two disadvantages of taking this drug in excess? (II)

e What is the reason why some persons are given morphia to take? (II)

f Although caffeine is a drug why does it seldom have serious effects? (II)

g Although amphetamines may have some benefit, what is the danger of such drugs? (II)

h What does the passage say is the reason for motorists being given a breathalyzer test? (II)

i How does drinking alcohol cause the kidneys to cease functioning properly? (II)

j Drinking alcohol on a cold day may make a person feel warmer but in fact it does not help them to maintain a constant temperature. Why is this? (II)

THEME 4 Development of organisms and continuity of life

4.1 A chip off the old block – variation and heredity

When you look at a large group of animals, for example penguins, it is very difficult to tell one from another. They all simply look the same. However, more careful observation will show small differences between them. Apart from identical twins, all living things vary. It is easier to see in humans probably because we are used to recognizing individuals. We learn this at a very early stage in our development. In the example given above, penguins learn to recognize each other but would find it difficult to tell one human from another. Even very closely related members of the same species vary in detail.

Fig. 4.1.1 A flock of penguins

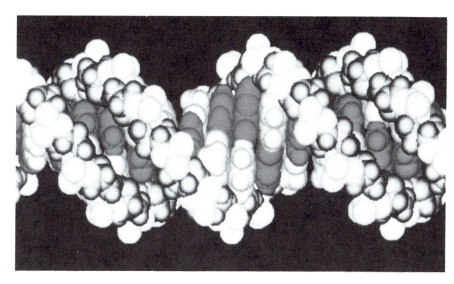

Fig. 4.1.2 Part of a DNA molecule

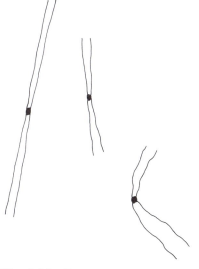

Most variations are inherited. People resemble their parents and have a mixture of features from their mother and father. The mixing of these features is possible because of **genes**. These are responsible for controlling all inherited features and are found arranged along chromosomes in every living cell. A gene is a number of active sections of a chemical called **DNA** (**deoxyribonucleic acid**) which are able to instruct the development of a particular characteristic (see Fig. 4.1.2). For example, genes control hair colour, eye colour, blood groups and behaviour. In order to understand how genes are passed from parent to offspring, it is necessary to understand something of how chromosomes behave.

Chromosomes are thread-like structures (see Fig. 4.1.3) which appear in the nuclei of cells when they divide. Every species has a fixed number of chromosomes in its cells. For example, humans have forty-six in every nucleus of every cell except for eggs and sperms. Of the forty-six, twenty-three have come from the father and twenty-three have come from the mother. Eggs and sperms have twenty-three chromosomes. When an egg meets a sperm during fertilization, their combined chromosomes make a total of forty-six again.

Fig. 4.1.3 Three chromosomes

When cells, other than eggs and sperms, form by cell division, they are identical in every respect. This means that each will have the same number of chromosomes and the same genes. It happens by a type of cell division called **mitosis** (see Fig. 4.1.4).

Stages in mitosis

1 When a cell is about to divide, chromosomes become visible in the nucleus. In the human body, 46 chromosomes exist in every nucleus of every cell which is capable of cell division.
2 A region called the **centrosome** divides into two, each one of which moves to opposite ends of the cell. Short **microtubules** radiate from the centrosomes. The nuclear membrane disintegrates.
3 A **spindle** of ray-like fibres of protoplasm appears between the centrosomes.
4 The centrosomes duplicate themselves to form pairs of **chromatids** and become arranged at the equator of the spindle.
5 The chromatids separate from each other, one member of each pair moving to the two poles of the nucleus.
6 The cytoplasm becomes constricted by the cell membrane.
7 The cell membrane between the new cells is completed. Nuclear membranes appear around each new nucleus. The chromosome material becomes dispersed once more. The mitotic stages will begin again when each cell becomes fully grown. Thus the body cells keep their characteristic number of chromosomes from generation to generation and therefore pass on genes from cell to cell.

Fig. 4.1.4   Stages in mitosis

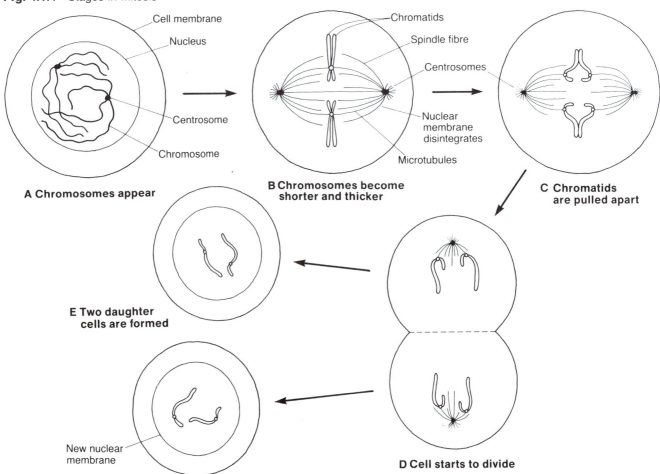

Cell membrane
Nucleus
Centrosome
Chromosome

A Chromosomes appear

Chromatids
Spindle fibre
Centrosomes
Nuclear membrane disintegrates
Microtubules

B Chromosomes become shorter and thicker

C Chromatids are pulled apart

E Two daughter cells are formed

D Cell starts to divide

New nuclear membrane

Stages in meiosis

In the previous account of the stages of mitosis it can be seen that the resulting cells possess the same amount of DNA as the parent cells. That is, 46 chromosomes in the case of Man.

The fusion of sex cells (eggs and sperms) during reproduction upsets this fixed number of chromosomes (and therefore the DNA), in the cell. For example, let us suppose that the number of chromosomes present in our cells is n. When we produce sex cells they will each contain n chromosomes. In the course of reproduction two sex cells fuse to form a **zygote**. The number of chromosomes in each zygote will then be $2n$, that is, n from each sex cell which led to its formation. Not only will there be twice as many chromosomes but these will be in pairs. Each chromosome will have a partner, identical in appearance, provided from the other sex cell. If we suppose that this zygote grows into an adult and produces its own sex cells with $2n$ chromosomes, then after fusion the zygote of the next generation will have $4n$ chromosomes, then $8n$, $16n$ and so on.

The problem then, is to avoid this multiplication of chromosomes. So at some stage the number of chromosomes must be halved to offset the effect produced by the fusion of the sex cells. This takes place by a special type of cell division, called **meiosis**. The stages are shown below. Note that, for simplicity, only 4 chromosomes are shown as the full number of chromosomes (**diploid** number). Therefore, in the example described above, $2n = 4$.

Fig. 4.1.5 Stages in meiosis

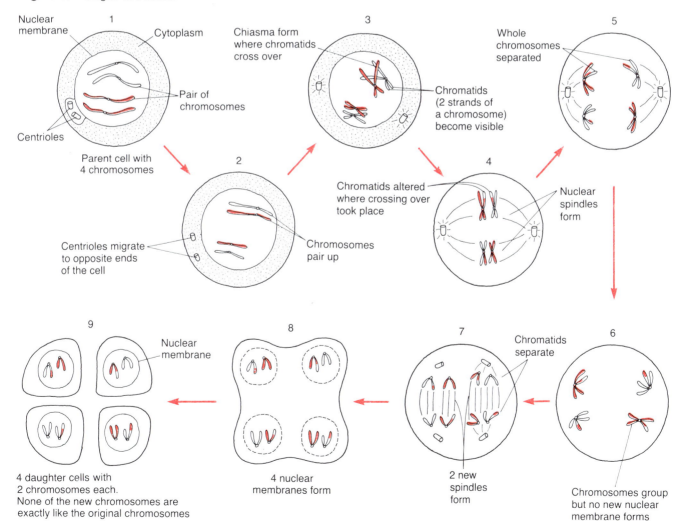

1 — Nuclear membrane; Cytoplasm; Pair of chromosomes; Centrioles; Parent cell with 4 chromosomes

2 — Centrioles migrate to opposite ends of the cell; Chromosomes pair up

3 — Chiasma form where chromatids cross over; Chromatids (2 strands of a chromosome) become visible

4 — Chromatids altered where crossing over took place; Nuclear spindles form

5 — Whole chromosomes separated

6 — Chromosomes group but no new nuclear membrane forms

7 — Chromatids separate; 2 new spindles form

8 — 4 nuclear membranes form; Nuclear membrane

9 — 4 daughter cells with 2 chromosomes each. None of the new chromosomes are exactly like the original chromosomes

Fig. 4.1.6

Genes – blueprints for our bodies

To build a human body requires a very detailed design plan provided by the genes in every cell of our bodies. A gene is defined as a 'unit of heredity'. These units linked together form the chromosomes. The scientific study of heredity is called **genetics**. It is scarcely a hundred years old but it has already made major contributions to the welfare of mankind. These range from improved crop plants and farm animals where selective breeding has greatly improved yields, to safe blood transfusion, which only became possible when the genetics of the blood group systems was understood.

Gregor Mendel, an Augustinian monk born in 1822, was the founder of genetics. He carried out cross-pollination experiments using different varieties of the garden pea plant, *Pisum sativum*. In this way he followed certain characteristics, such as size, seed colour, seed shape and flower colour, through several generations. Mendel was ahead of his time in his interpretation of the results of his experiments. He formulated a theory to explain the transmission of characteristics from one generation to another, a process relevant to every living organism.

Mendel cross-pollinated pairs of plants which were different in one of the chosen characteristics. An example is that of seed colour. By crossing a yellow-seeded plant with a green-seeded plant, he obtained a **first filial (F1) generation** of yellow-seeded plants. All the F1 plants had the characteristics of one parent – they were all yellow-seeded.

Mendel went on to self-pollinate the plants (to fertilize each plant with its own pollen) from the F1 generation. He found that some of the resulting **F2** plants had yellow seeds and others had green seeds, in the ratio of 3 : 1. Why was one of the original parent's green-seeded characteristics being expressed now, after missing a generation? He suggested that the plants of the F1 generation were **hybrids**, that they contained factors for both colours, but showed only one of them, the one that was **dominant**. Therefore, in the pea plant, we know that yellow-seed colour is dominant and green-seed colour is **recessive**.

Mendel went on to self-pollinate the plants of the F2 generation. The green-seeded plant reproduced its own variety. One of the yellow-seeded plants also reproduced its own variety, but the other two gave both yellow and green-seeded types in the ratio of 3 : 1 as in the F2 generation (see Fig. 4.1.7).

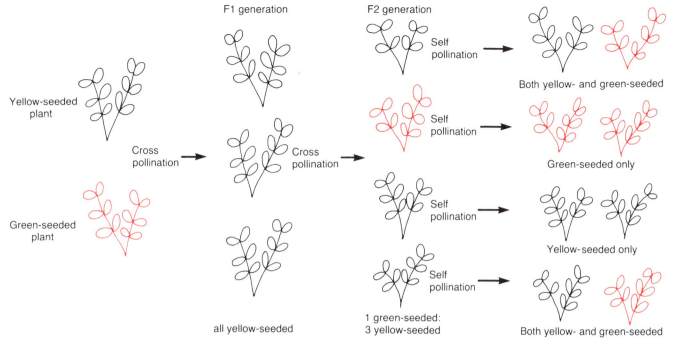

F1 generation

F2 generation

Yellow-seeded plant

Green-seeded plant

Cross pollination

Cross pollination

Self pollination — Both yellow- and green-seeded

Self pollination — Green-seeded only

Self pollination — Yellow-seeded only

Self pollination — Both yellow- and green-seeded

all yellow-seeded

1 green-seeded: 3 yellow-seeded

Fig. 4.1.7 Mendel's experiments

The explanation suggested by Mendel showed his great ability to interpret the results and a surprising insight into the mechanism of heredity. Mendel stated that for every **specific characteristic**, each organism possesses two factors which may be the same (**homozygous state**) or different (**heterozygous state**). In the formation of the **gametes**, these factors segregate independently of one another, and only one of the pair, for instance yellow or green seed colour, is passed on to the offspring. The choice of factor to be transmitted from each parent is random. One factor is usually dominant, so if an organism is heterozygous only that one will show.

Although Mendel published his work in the middle of the nineteenth century, its importance was not realized until his work was rediscovered years after his death.

Rapid progress has been made this century in the investigation of heredity. Very important steps forward were the discovery of the structure of the key molecule, DNA, and the method by which protein is synthesized. We now know that the factors which Mendel spoke of are pieces of DNA. They have been given the name genes, and the different forms in which one gene can exist, such as those producing yellow or green characteristics in pea seeds, are called **allelomorphs** or **alleles**.

If a person is homozygous for a certain characteristic he must have the same form of the gene representing that characteristic, the same allele, on both chromosomes of the **homologous** pair (a pair of chromosomes, the partners of which have come from the mother and father respectively). If he is heterozygous, the alleles will differ.

Looking at the F2 generation, we saw that three-quarters of them were yellow-seeded plants, but then on further self-pollination one-third of these produced all yellow-seeded plants, showing that they were homozygous for the yellow-seed allele, while two-thirds gave the 3 : 1 ratio obtained in the F2 progeny, showing they were heterozygotes.

Another two terms used in genetics can be applied here. All the yellow-seeded plants looked the same so they were **phenotypically** similar, but two different **genotypes** were involved. (A genotype is the genetic constitution of an organism.)

How proteins are made

So far genes have been described as being linked together to form chromosomes in the nucleus of every living cell. Now their structure will be more precisely defined. A gene is a molecule of DNA, a nucleic acid, which is formed by two strands twisted around each other in a spiral. Each strand contains, among other chemicals, four types of chemicals called **bases**. These bases form bonds between the strands and always pair in the same combinations (see Figs. 4.1.8 and 4.1.9). The bases are: **adenine (A)**, **guanine (G)**, **cytosine (C)**, and **thymine (T)**. They pair A with T, C with G and also A with **U** (**uracil**), a base that replaces T in **RNA** (another type of nucleic acid).

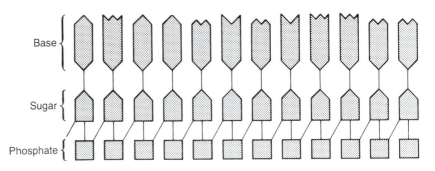

Fig. 4.1.8   A model of a single strand of DNA

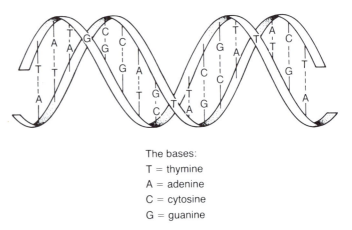

The bases:
T = thymine
A = adenine
C = cytosine
G = guanine

Fig. 4.1.9   The spiral nature of DNA

Three of these bases together form a code symbol, known as a **codon**, for the synthesis of a specific **amino acid**. Each **protein** is made up of chains of amino acids in a certain order which are called **peptides** or, if slightly longer, **polypeptides**.

The function of genes is to direct the production of the proteins required by the cell throughout its life. Protein synthesis requires two main processes, first transcription of the information contained in the gene's structure onto a messenger, and then translation of the messenger's information into the correct sequence of amino acids to create the protein which is needed (see Fig. 4.1.10).

The DNA in each gene acts as a **template** (pattern) for the specific **messenger RNA** (**m-RNA**) molecule. The production of this new strand of nucleic acid, m-RNA, depends on the fact that the bases pair with their specific partners, so that DNA containing a certain sequence of bases produces only one specific sequence of bases in the m-RNA molecule which comes off its template. The new molecule of m-RNA migrates through the nuclear membrane into the cytoplasm of the cell where ribosomes lie.

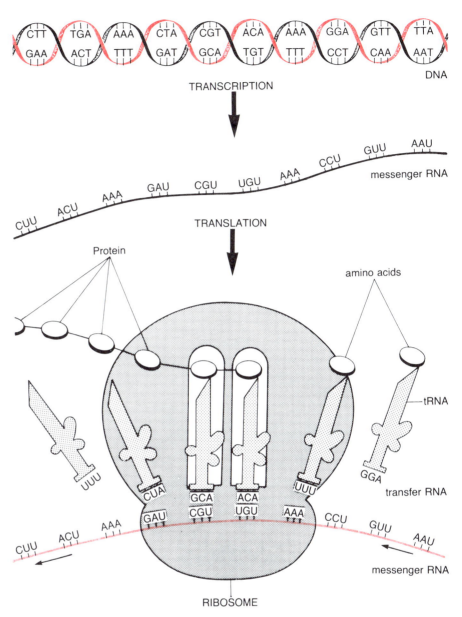

Fig. 4.1.10 Protein synthesis

Ribosomes contain a mixture of ribosomal RNA and protein. Once attached to the ribosome, the m-RNA in turn acts as a template from which its information is translated. A third type of ribonucleic acid, **transfer RNA** or **t-RNA**, occurs in the cytoplasm. Each t-RNA molecule contains only three bases, so it codes for only one specific amino acid. The m-RNA relays its coded message to the t-RNA molecules which then move through the cytoplasm to pick up their respective amino acids.

The resulting complexes return to the m-RNA ribosome unit and are attached one by one to the other amino acids in the order dictated by the m-RNA. Once the t-RNA has deposited its amino acid in the correct position in the peptide chain, it is free to operate again. In fact, a t-RNA molecule can carry out this process three times every second.

We have come a long way from Mendel and his experiments with garden peas. Although it is interesting to follow the transmission of genes for physical characteristics, such as height, tongue rolling, hair and eye colour (see Fig. 4.1.11), the main interest in human genetics is in hereditary disease. Research is being carried out to attempt to eliminate **mutant** (abnormal) genes from our population.

Fig. 4.1.11   The transmission of genes for some physical characteristics (after Ann Morrow, Camera Press)

Biochemical studies on affected people may reveal which protein (enzyme) is defective and bring the possibility of treatment nearer. Genetic registers are being set up to accumulate information about families in which there is a mutant gene. Members of the family who are at risk of being affected or of passing on the abnormality can receive genetic counselling, and may then decide to restrict their family size.

Disorders in chromosomes and genes

Fig. 4.1.12 A Down's syndrome child

Down's syndrome
(see Fig. 4.1.12)

Down's syndrome is caused by the presence of an extra chromosome, i.e. 47 instead of 46 in every living cell (see Fig. 4.1.13). The condition is named after Dr Down who first recognized it.

Affected children are short for their age and are mentally retarded. The head is smaller than normal, the face is round, the eyes slanted, and there is a fold of skin across the inner corner of the eyes. The hands are short and stubby, often showing a curved little finger. The fingerprints and palm prints are also abnormal and the feet have a wide gap between the first and second toes. Internal problems may lead to heart disease and abnormal development of the intestine.

The internal abnormalities and an above-average tendency to contract infectious diseases used to be the cause of early death of many children with Down's syndrome. With the advance of medical knowledge the outlook for survival has improved greatly. Although needing care in a sheltered environment, the patients can live happy lives and are exceptionally lovable people.

Down's syndrome is one of the most common causes of mental handicap, affecting almost one baby in every 600 born. The risk of having an affected child depends on the age of the mother. For women in their twenties, the figure is about 1 in 2000; the risk rises at the age of 32 and reaches a maximum of 1 in 50 for women of 45 or more.

The extra chromosome can come from either parent. It comes from a mistake in the division of the chromosomes during the formation of either the egg or the sperm. The defect is therefore present even before the egg has been fertilized.

Fig. 4.1.13 Chromosomes of a Down's syndrome child

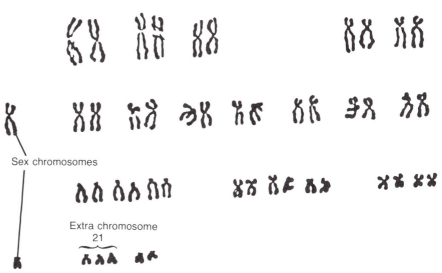

A gene disorder – sickle-cell anaemia

This is an often fatal disease which is quite common in West Africans. If there is a low level of oxygen in the blood, red blood cells of a person suffering from this disorder collapse into a sickle shape and may form blockages in blood vessels.

The disease is inherited as a single mutant (freak) recessive gene. If a child inherits the gene from both parents it has only a 20 per cent chance of surviving into adulthood. The gene that controls the formation of haemoglobin is not formed properly. Therefore the haemoglobin in the sickle-shaped red cells is not normal and is not very good at carrying oxygen.

If this is so important you might think that, over thousands of years, the mutant gene would have disappeared from human chromosomes because carriers would die before having children. However, under certain circumstances, it is an advantage to have some sickle-shaped red blood cells in your blood system (Fig. 4.1.14). The reason is because the malarial parasite is less likely to attack people who are heterozygous for the mutant gene.

For example, let us assume that the gene for haemoglobin is Hb and that the gene for sickle-cell haemoglobin is HbS. Then a normal person will have the haemoglobin genotype Hb Hb. A person with the sickle-cell disorder will have the genotype HbS HbS and usually dies. However, a person with the genotype HbS Hb can still survive and will be resistant to malaria.

If two people, each with a genotype HbS Hb, have a child then the predicted genotype of the child can be shown as follows:

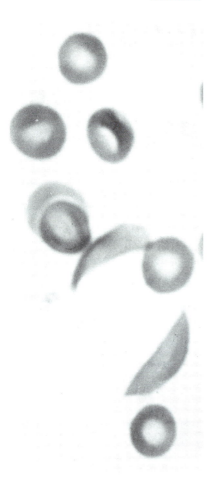

Fig. 4.1.14   Photomicrograph of sickle-cells

Parents HbS Hb × HbS Hb

♀ gametes ♂	HbS	Hb
HbS	HbS HbS	HbS Hb
Hb	Hb HbS	Hb Hb

It can be seen that there is a one in four chance of the child dying with the sickle-cell disorder (HbS HbS). There is a one in two chance of the child being a carrier of the sickle-cell gene but probably surviving (HbS Hb). There is a one in four chance of the child being perfectly normal for the haemoglobin gene (Hb Hb).

The common errors – cystic fibrosis

One of the most common genetic disorders is **cystic fibrosis**. It is an inherited disease which affects the pancreas and the bronchioles of the lungs. It is one of the most common fatal diseases of childhood and is inherited as a recessive gene in the following way:

Let C = the gene for a normal pancreas and bronchioles.
Let c = the recessive gene for cystic fibrosis.

A person suffers from the disease only if he or she has two genes for cystic fibrosis, i.e. the genotype cc. A person with the genotype Cc is a carrier of the disorder but does not suffer from the disease.

There is therefore a one in four chance of a child suffering from the complaint if two carriers have a child.

♀ gametes ♂	C	c
C	CC	Cc
c	cC	cc

Carrier man Cc — Carrier woman Cc

What would be the offspring's genotype of a cross between a normal man and a woman who suffers from cystic fibrosis?

Phenylketonuria A certain enzyme is not able to be made in people suffering from this condition. This enzyme is responsible for changing a chemical called **phenylalanine** to an amino acid used by the body. Phenylalanine occurs in meat, fish, cheese, eggs, wheat and butter. In the absence of the enzyme which acts on it, phenylalanine accumulates in the tissues. Some is converted to a dangerous chemical which causes damage to the central nervous system. It leads to a person being mentally retarded and is due to a single abnormal gene. About one case in 15 000 babies suffers from the problem and synthetic substitutes for protein must be given in the diet.

Sex linkage Some disorders that are inherited occur much more often in men than in women. Two well known examples are colour-blindness for red and green and haemophilia (see p. 146). These disorders are due to **sex-linked genes**. Since the X chromosome is larger than the Y (see p. 222), it contains some genes that are not present on the shorter Y chromosome.

The most common sex-linked genes are recessive. For example, red-green colour blindness is recessive (c), normal colour vision is dominant (C). Therefore, if a woman has one normal X chromosome (C) and one X chromosome with the defective gene (c), she will be a carrier of the condition (Cc). The dominant normal C is expressed. However, if a male has an X chromosome with the recessive gene (c), he will be colour blind for red and green. This is because the other chromosome, Y, has no gene to pair with c. With no normal colour vision gene as a dominant character, the recessive gene will express itself. Haemophilia is inherited in exactly the same way.

$X^C Y$ — Normal father $X^C X^c$ — Carrier mother

♀ gametes ♂	X^C	X^c
X^C	$X^C X^C$	$X^C X^c$
Y	$X^C Y$	$X^c Y$

half the female children will be normal $X^C X^C$
half the female children will be carriers $X^C X^c$
half the male children will be normal $X^C Y$
half the male children will be red-green colour blind $X^c Y$

Is it theoretically possible for a female to be colour blind for red-green? Explain why.

Multiple genes   In this type of inheritance, certain genes have equal dominance to each other. One of the most common examples is seen in blood groups (see p. 141). The inheritance of a particular blood group is determined by a gene that controls the production of an antigen.

Group A has A antigen. Group B has B antigen. Group AB has both antigens. Group O has neither antigen.

A and B have equal dominance to each other; both are dominant over group O. The possible genotypes of blood groups are shown in the table:

Phenotypes (blood group)	A	B	AB	O
Possible genotypes	AA or AO	BB or BO	AB	OO

If we know whether the parents are heterozygous or homozygous, we can find out the possible blood groups of their children.
Example:

Consider two parents: Mother group B   Father group A

There are four sets of possible genotypes:

(1)		(2)		(3)		(4)
BB × AA	or	BO × AO	or	BO × AA	or	BB × AO
mother father		mother father		mother father		mother father

The possible offspring would be produced as follows:

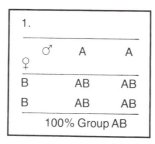

1.

	♂ A	A
♀ B	AB	AB
B	AB	AB
100% Group AB		

2.

	♂ A	O
♀ B	AB	BO
O	AO	OO

25% Group AB
25% Group A
25% Group B
25% Group O

3.

	♂ A	A
♀ B	AB	AB
O	AO	AO

50% Group AB
50% Group A

4.

	♂ A	O
♀ B	AB	BO
B	AB	BO

50% Group AB
50% Group A

Questions: Domains I and II

1

a Hair colour in humans is an inherited characteristic. Black hair is determined by a dominant gene B; red hair is determined by a recessive gene b.

In the family shown in the diagram, Mr Smith has black hair and Mrs Smith has red hair. John and Susan have black hair; Peter and Debra have red hair. Using the symbols B and b, complete the diagram to show the genetic make up of the family. (II)

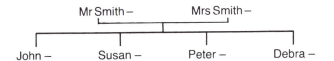

b Which of the children are heterozygous for hair colour? (II)

c Explain why Mr Smith could not have two genes for black hair. (II)

d If Mr and Mrs Smith had another child, what are the chances of that child having red hair? (II)

e Which of the following are inherited: language; sex; diphtheria; blood group? (I)

f Explain the meaning of the term 'mutation'. (I)

g Give one example of a mutation. (I)

3 The diagram shows some of the genetic features of a marriage between a man who has haemophilia and a woman who is a carrier for this disease, and the possible children who could be produced.

Normal clotting is controlled by a dominant gene (H), the inability of blood to clot (haemophilia) is controlled by a recessive gene (h). These genes are present on the sex chromosomes.

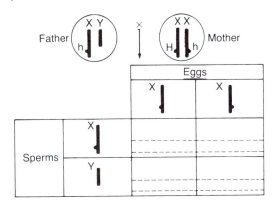

Copy the diagram and complete it by:

i showing the gene for clotting and haemophilia present in the sperms and eggs;

ii showing the sex chromosomes present in each of the four possible offspring;

iii showing the genes present in the offspring;

iv showing the sex of each offspring;

v showing for each offspring whether it has normal clotting or haemophilia. (II)

2 The diagram shows a pedigree involving the Rhesus factor.

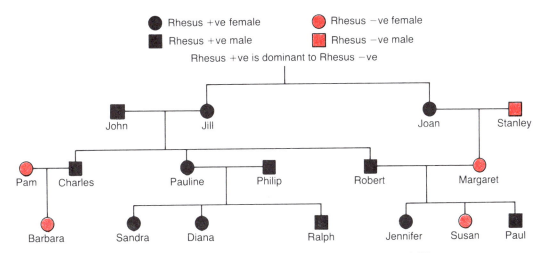

a What is the family relationship between:
  i Pauline and Margaret?
  ii Joan and Charles? (II)

b What is the Rhesus genotype of each of the following?
  i Stanley?
  ii Charles? (II)

c Give reasons for your answer. (II)

d If Jennifer and Ralph were to marry, could any of their offspring have Rhesus negative blood? Give a reason for your answer. (II)

4 The family pedigree in the diagram shows four generations of a family which has a history of inherited blindness called optic atrophy. The blindness is inherited as a sex-linked recessive gene on the X chromosome.

a From which of their parents do the children inherit the blindness? (II)

b To which of their children do the blind parents pass on the condition?

c Record the sex chromosomes and the genes for this condition carried by
  i a blind male;
  ii a normal male;
  iii a daughter of a blind male. (II)

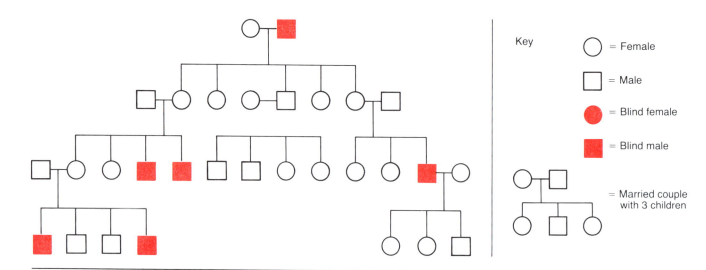

Key

○ = Female

□ = Male

● = Blind female

■ = Blind male

= Married couple with 3 children

5 Sickle-cell anaemia is caused by a mutant gene HbS. The normal gene is Hb. The homozygous condition of the mutant gene is lethal but it is possible for the heterozygotes to survive.

a Show as a punnett square the offspring of a cross between:
  i a normal parent and one carrying the sickle-cell gene;
  ii two parents, both carrying the sickle-cell gene. (II)

b The percentages of heterozygotes occurring in different areas are shown in circles on the map of parts of West Africa. The distribution of the malarial parasite is also shown.
  i Explain the distribution of the sickle-cell gene and that of the malarial parasite.
  ii Mosquitoes carry the malarial parasite. Suggest an explanation for the distribution of the malarial parasite shown on the map. (II)

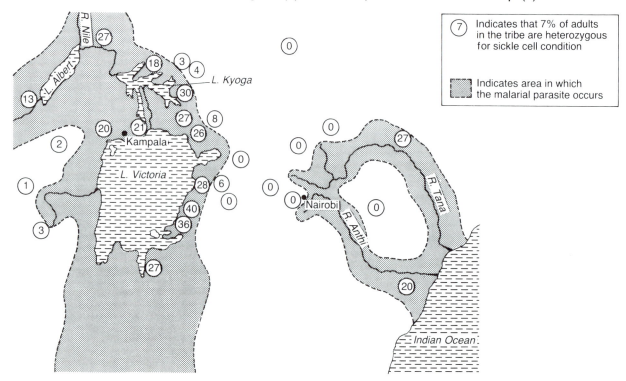

⑦ Indicates that 7% of adults in the tribe are heterozygous for sickle cell condition

Indicates area in which the malarial parasite occurs

6 The table shows some of the characteristics of five pupils in a class.

Name	Height (cm)	Hand span (mm)	Eye colour	Blood group	Left/right handed
Bethan	149	169	Blue	O	Right
Susan	150	165	Blue	A	Right
Peter	152	172	Brown	B	Right
Monica	154	172	Grey	O	Right
Mary	151	170	Blue	AB	Left

a Who is (i) left handed;
      (ii) the tallest? (II)
b Who has the shortest hand span? (II)
c How many pupils have blue eyes? (II)
d Which pupil is the shortest? (II)
e Who has grey eyes? (II)
f What percentage of pupils are right handed? (II)
g Height is an example of continuous variation. Which other feature in the table is also an example? (I)
h Which two features are due only to inheritance? (I)

7 Complete the diagram to show the genotypes with respect to ABO blood grouping of the parents, gametes and offspring. State the blood group of each child. (I)

	Mother's blood group – heterozygous A	Father's blood group – heterozygous B
Genotype	———	———
Gametes	———	———
Genotype of children	———	———
Blood group of children	———	O

8 The heights of 34 sixteen-year-old boys were measured in cm.

168	163	166	169	171	173	158	159	172	167	154	172	165
158	160	162	166	151	174	168	170	163	174	166	164	170
169	157	168	161	167	169	165	167					

a Sort out the results and with the aid of the grid below, indicate how many heights fall into each division.

Heights range (cm)	150–154	155–159	160–164	165–169	170–174
Numbers					

(II)

b Plot a histogram on graph paper to show the range of heights in this group of boys. (II)
c State one other characteristic which could have been chosen to show the same kind of variation. (I)
d How would the graph differ for girls of the same age? (I)
e Give a reason for your answer. (I)

4.2 The life machine

The ability to reproduce is essential to the survival of our species. The human **reproductive system** is probably the most advanced in the animal kingdom. We can think of it as a machine which produces members of the next generation. The machine has two quite separate parts, **male** and **female** (see Figs. 4.2.1 to 4.2.4).

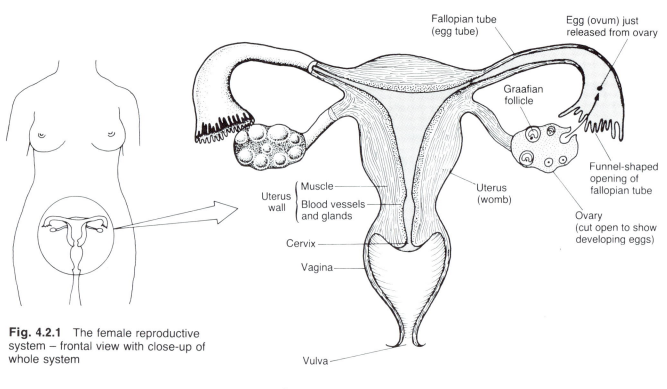

Fig. 4.2.1 The female reproductive system – frontal view with close-up of whole system

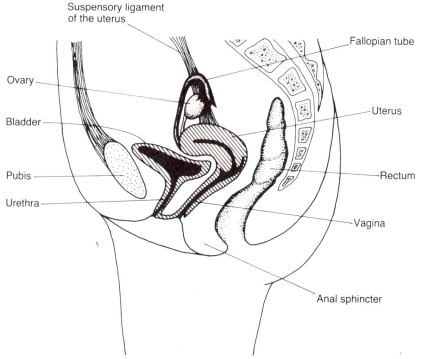

Fig. 4.2.2 Side view of female reproductive system

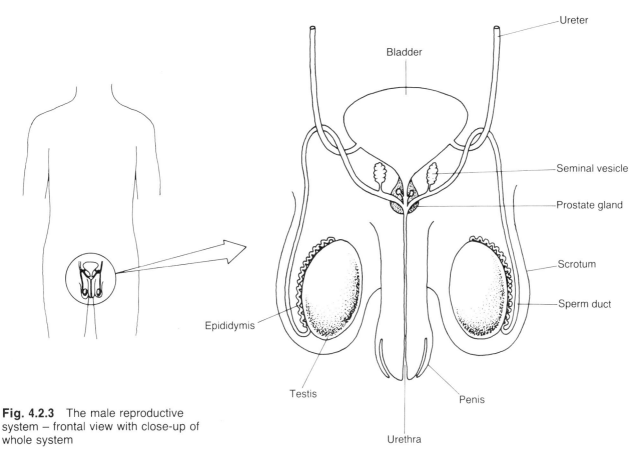

Fig. 4.2.3 The male reproductive system – frontal view with close-up of whole system

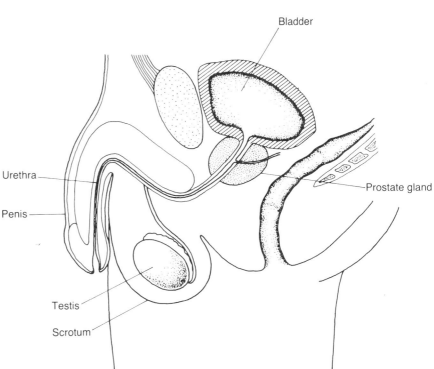

Fig. 4.2.4 Side view of male reproductive system

There is a fusing of some parts, the **sperm** and the **egg** to produce a new product. This is the **fertilized egg** or **zygote**. The zygote is then allowed to mature and develop on the production line – the **fallopian tube** and the **uterus**. After this, it leaves the line by the process of childbirth.

Of course any manufacturing process needs to have a set of blueprints for the finished product. In this case the manufacturing instructions are contained within the genetic material carried inside the egg and sperm (see Figs. 4.2.5 and 4.2.6). In addition the machine needs to be internally controlled. In the case of the reproductive machine much of this control is by chemicals called **hormones**.

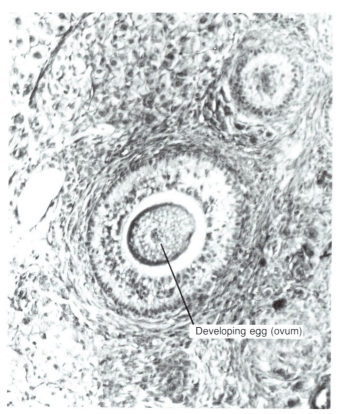

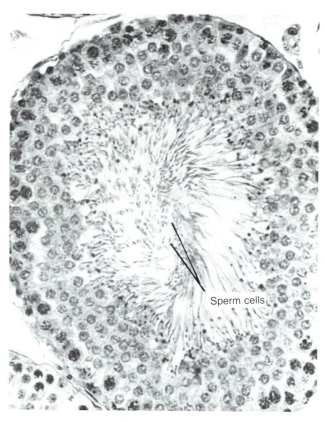

Fig. 4.2.5 Photomicrograph of an egg cell

Fig. 4.2.6 Photomicrograph of sperm cells

The human reproductive system differs from that of many other animals in certain basic principles. The first is that in all mammals fertilization is internal within the body of the female. Internal fertilization ensures a greater degree of protection for the zygote than fertilization that occurs outside the female body.

Mammals also differ from many other forms in that their young are generally born capable of a separate existence rather than being laid as fertilized eggs. In the human, the **foetus** (developing baby) stays within the womb (uterus) until much of its physical development is completed. After birth it still requires a long period of care and feeding before it can look after itself. This is a longer period than is found in many other mammals. Our childhood is a time of great importance because we use it to learn about our environment from our more experienced parents.

Fully-equipped at birth   The unfertilized human egg is a spherical single cell about one tenth of a millimetre in diameter. This is a very large cell, when compared with most cells in the body. It has a nucleus, containing long, thread-like chromosomes (see p. 203) on which is coded the genetic information for the development of the future generation. Most of the rest of the egg is concerned with feeding the zygote as it grows and develops.

The eggs of mammals grow, develop and mature in the **ovaries**. These are each about the size of a walnut and are inside the abdomen of women.

A girl is born with all the eggs she will ever have. She is assured of all she will want throughout her reproductive life. After the age of twelve, she will release about 400 at the rate of one every month. Each egg completes its development in the part of the ovary called the **Graafian follicle**. When the egg leaves this it is carried into one of the paired fallopian tubes that run to the uterus.

A look at the sperm

In the great majority of cases, fertilization occurs within the fallopian tubes. Usually, several thousand sperm are present in the tube following intercourse but only one unites with the egg. A sperm consists of a flattened paddle-shaped head and a long movable tail which drives the sperm forwards (see Fig. 4.2.7). The length of a sperm is only half the diameter of an egg. During intercourse a male passes about 300 million sperms into the female **vagina** from his **penis**.

Only a few thousand sperm move through the entire female reproductive tract to the fallopian tubes. From the vagina they pass through the **cervix** (neck) of the uterus before entering the fallopian tube.

The sperms are produced in the **testes**. Within each testis there are a large number of very fine hollow tubes with walls lines by special cells. These cells produce sperm that are shed into the tubes (see Fig. 4.2.8). All the genetic material in the cells is contained in the sperms' heads.

The fine tubes all connect within each testis and empty into a larger tube called the **epididymis** which runs down the side of the testis. During the passage through the epididymis the sperms become mature. From the epididymis they enter a thick-walled **vas deferens** – yet another tube. The vas deferens leads to the **urethra** which is a tube that carries both the sperms and urine through the penis. At the base of the vas deferens is the **prostate gland**. This makes a liquid in which the sperms can swim towards the female egg cell. The liquid plus the sperms is called **semen** or **seminal fluid**.

Fig. 4.2.7 Photomicrograph of sperms

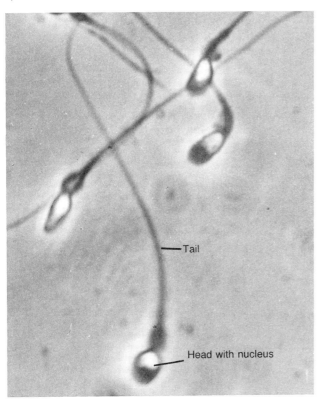

Tail

Head with nucleus

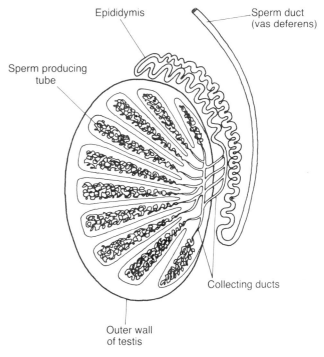

Epididymis

Sperm duct (vas deferens)

Sperm producing tube

Collecting ducts

Outer wall of testis

Fig. 4.2.8 The testis

A girl or a boy?   During the act of intercourse, the penis becomes erect due to the pressure of blood in a spongy tissue within it. Erection of the penis is an automatic response to sexual arousal in preparation for intercourse and fertilization.

If one sperm succeeds in penetrating an egg, fertilization occurs (Fig. 4.2.9). The nuclei of the egg and sperm actually fuse together. The resulting zygote will contain the genetic material of both the father and mother. All normal women have in the cells of their bodies 22 pairs of normal chromosomes (**autosomes**) and one pair of X chromosomes (**sex chromosomes**).

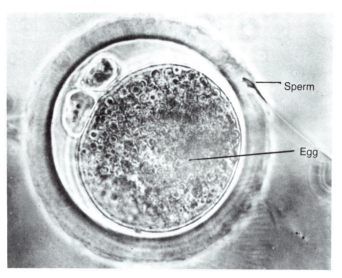

Fig. 4.2.9   Photomicrograph of fertilization

In men, no such pair of X chromosomes exists. Each cell contains one X chromosome and a smaller partner called the Y chromosome. If a zygote contains a Y chromosome it will develop into a boy. In its absence the baby will be a girl (see Fig. 4.2.10).

The egg cell has only one of each chromosome pair in its nucleus. So it has one X chromosome. By contrast, a sperm cell has either an X or a Y (also one of the original pair). If an X sperm fertilizes an egg, all the cells that develop from the fertilized egg must be female because they are XX. In the same way, a Y sperm will give a zygote with the make-up XY, which will develop into a boy (see Fig. 4.2.10). The sex of the baby therefore depends on the type of sperm that fertilizes the egg. Since X and Y sperms are equally abundant, the ratio of boys to girls should be roughly equal. A study of populations shows that this is so.

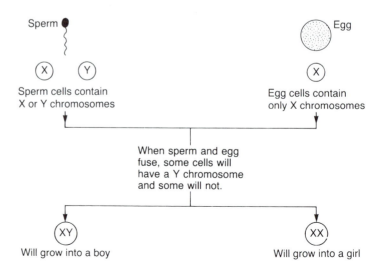

Fig. 4.2.10   Sex determination

After fertilization the zygote starts to divide, first into two cells, then four and so on (see Figs. 4.2.11–4.2.13). As it does so, it moves down the fallopian tube and enters the uterus. Here the zygote becomes a hollow ball of cells called the **bastocyst**. Seven days or so after fertilization this **implants** itself in the wall of the uterus. As a result of this implantation, the fertilized egg comes in close contact with the mother's blood supply from which it obtains its food and oxygen. A new organ, the **placenta**, is formed between the mother and the foetus (see Figs. 4.2.12 and 4.2.13).

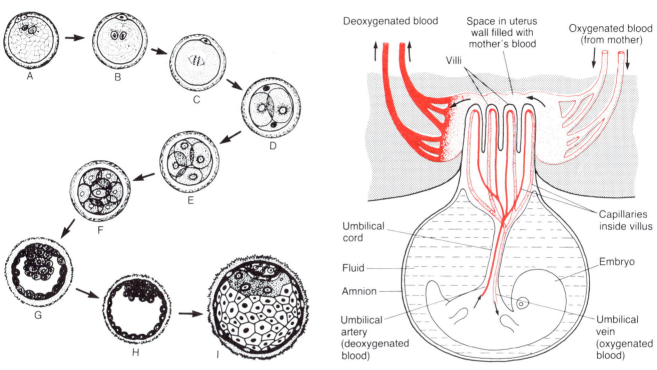

Fig. 4.2.11   Stages in the division of the zygote to form the blastocyst

Fig. 4.2.12   The formation of the placenta

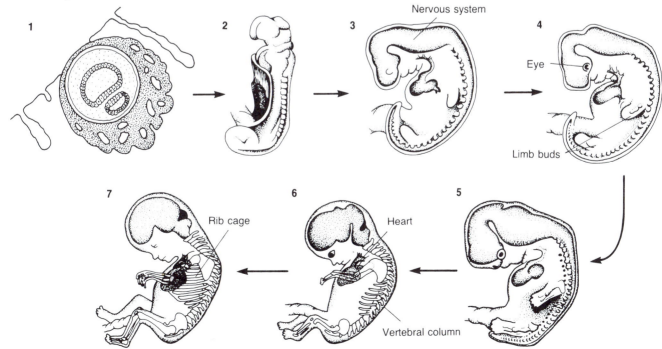

Fig. 4.2.13   Stages in the development of the embryo (after Yvonne Banks, Orbis)

Control   The reproductive system in men and women is under the control of the brain and hormones. Both types of control are closely connected and ensure the normal functioning of the system. The ovary produces two female sex hormones called **oestrogen** and **progesterone**. These account for the characteristic signs of femaleness. At puberty these hormones are responsible for breast development and for the distribution of fat under the skin. Together these produce the characteristic form of a woman's body. During pregnancy, these and other hormones are also very active. Without them, pregnancy would not be possible. These hormones also maintain the menstrual cycle.

From the age of puberty until the time when eggs are no longer released from the ovaries (the **menopause**) most women show a regular pattern of bleeding from the vagina. It occurs roughly every 28 days and is called **menstruation** (Fig. 4.2.14). The bleeding is caused by a breakdown of the lining of the uterus (**endometrium**). It passes out of the uterus through the cervix and appears in the form of menstrual blood.

The time between menstrual bleeding can be divided into two phases. In the first, the endometrium is being built up. It is under the influence of the hormone oestrogen. During this phase a new egg cell is developing inside the ovary. About half-way between menstrual periods egg release (**ovulation**) occurs.

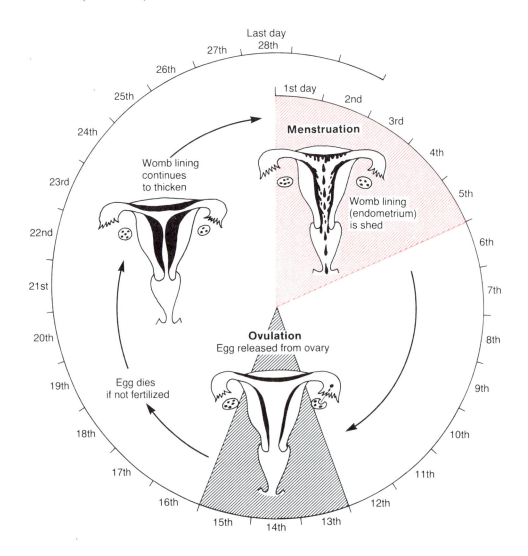

Fig. 4.2.14 The menstrual cycle and ovulation

After the egg has left the ovary, hormones continue to be produced. During this phase the endometrium is stimulated by both oestrogen and progesterone. The Graafian follicle, which produced the egg, changes into a gland called the **corpus luteum** (see Fig. 4.2.15). If no pregnancy occurs then this gland only remains for eight to ten days. Then it breaks down and for a while hardly any oestrogen or progesterone are secreted. In the absence of these hormones the lining of the uterus breaks down and is shed in the menstrual blood.

At the end of menstruation a new follicle starts to develop in the ovary. Oestrogen is again secreted. The endometrium is built up again and another menstrual cycle starts.

The brain's main part in the control of the reproductive system occurs through its pituitary gland which is the master hormone gland of the body. It is called this because it produces hormones which control the development of all the other hormone-producing glands of the body. It produces **gondotrophins** which regulate the production of oestrogen and progesterone (see p. 190).

Reproduction in men is also under the control of hormones. The main one is the sex hormone called **testosterone**. It is made in special cells in the testes and is responsible for maintaining sperm production as well as producing the secondary sexual characteristics of males. These include breaking of the voice, growth of the penis and growth of facial and pubic hair.

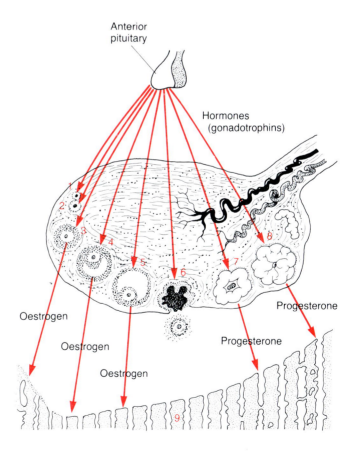

Anterior pituitary

Hormones (gonadotrophins)

Oestrogen

Oestrogen

Oestrogen

Progesterone

Progesterone

1 – 5 (Stages in the development of a graafian follicle)

6 (Ovulation)

7 – 8 (Formation of the corpus luteum)

9 (Stages in the thickening of the lining of the uterus)

Fig. 4.2.15 Hormonal action during the menstrual cycle

The journey towards birth

The nine-month growth of the tiny fertilized egg towards the newborn child is called **gestation** and is the most eventful period of change that the body experiences. A baby begins its life as the fertilized egg called a zygote. It is formed by the union of sperm and egg at the moment of fertilization.

Within a cell of about one ten-thousandth of a centimetre in diameter is all the information needed to change the zygote into the highly-complex newborn child of about 50cm in length. There is a great increase in the number of cells. They are produced at rates which vary according to where they are within the embryo. They change according to their future roles. The fertilized egg contains all the information necessary to bring this about.

We can divide the period of growth of the baby into three stages: the **pre-embryonic**, the **embryonic** and the **foetal**. The pre-embryonic stage lasts three weeks and includes repeated division of the zygote to form a large balls of cells, the blastocyst (see p. 223).

The embryonic stage lasts until the end of the eighth week during which all the internal structures are laid down. At the end of the embryonic stage the **umbilical cord** is properly developed, a tail is present and the eyes are beginning to appear. The embryo is now about 30mm long. The brain and head continue to grow rapidly and the face is formed.

Fig. 4.2.16   Stages in the growth of the embryo and foetus

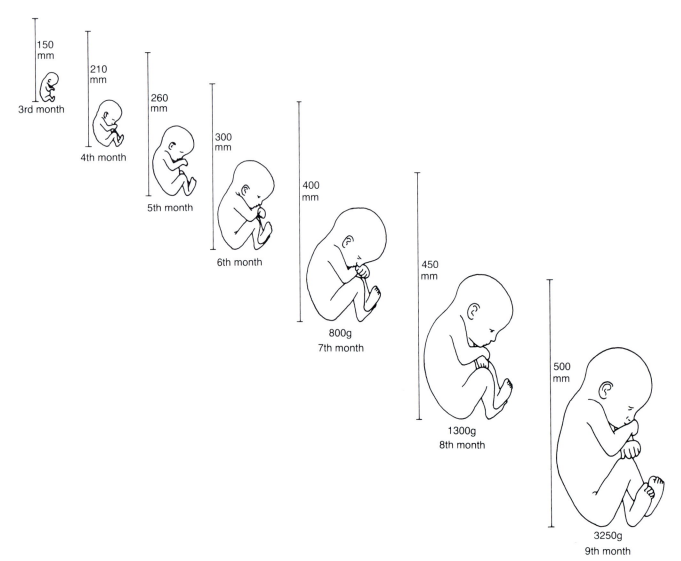

150 mm
3rd month

210 mm
4th month

260 mm
5th month

300 mm
6th month

400 mm
800g
7th month

450 mm
1300g
8th month

500 mm
3250g
9th month

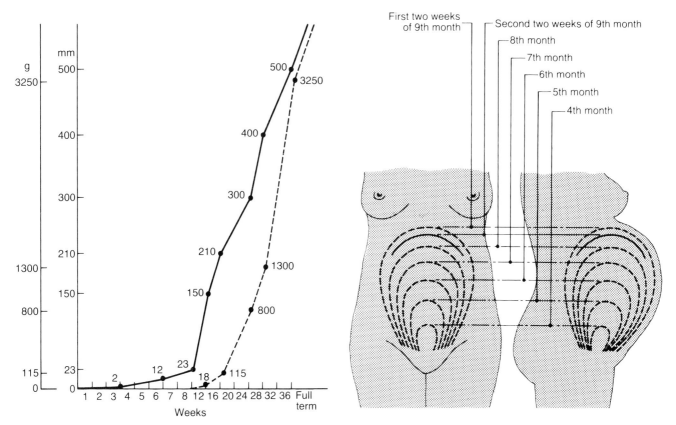

Fig. 4.2.17   Graph showing growth rate of the embryo and foetus

Fig. 4.2.18   The stages in the growth of the womb

At the same time the tail gets smaller and by the end of the eighth week, the embryo begins to look human. See Figs. 4.2.16 to 4.2.18.

The foetal period lasts from the beginning of the ninth week until birth and mainly involves growth and development of structures already present rather than the formation of new ones. During the third month growth is still rapid and the foetus almost doubles in length. Throughout the foetal period the head is large and the body and legs are relatively small. Early in the third month the eyelids develop and hair starts to appear. By the end of the third month hair has formed as a very fine coat called **lanugo**. The limbs can now be seen as human arms and legs and the fingernails and toenails are beginning to appear. By the fourth month the length of the foetus is about 150mm, excluding the legs, as in a sitting position. At five months the foetus starts to move for the first time. By the sixth month the features are more like those of a newborn baby. The lanugo is becoming darker and the skin is very wrinkled. This wrinkled effect is lost by the seventh month as fat begins to be laid down under the skin. From this stage onwards alveolus formation in the lungs begins. This process continues until after birth.

The late development of the alveoli is one of the reasons why **premature** babies (those born before full development) are at a disadvantage. Their respiratory system is not fully formed. Premature babies of this age will often survive if very carefully treated.

During the eighth and ninth months the hair on the head becomes more obvious than the lanugo, which begins to disappear. Fat continues to be laid down and growth is rapid. When the baby is ready to be born it weighs about 3250g and has an overall length of about 500mm. It is about to begin its existence in the world outside the womb. But the most decisive nine months of its life already lie behind it.

Childbirth   During the first three months in the womb, the embryo is protected in a bag of water, (the **amniotic fluid**), which forms like a balloon round the embryo, which floats in the fluid. The contents of the womb are sealed off by a plug of mucus in the cervix, so the baby is contained within its own sterile warm and shockproof capsule. From the third month onwards, the placenta functions to transport oxygen and food to the developing foetus. It also carries the foetal waste products back to the mother's bloodstream. The foetal and mother's bloodstreams do not actually mix. The placenta contains a fine membrane through which chemicals are sieved off. The placenta consists of a delicate branching network of blood vessels and is essential for the foetus's continued growth.

Towards the end of **gestation** the baby moves its position and goes down into the bony pelvis with its head like an egg in an egg cup (Fig. 4.2.19). It has already started on its journey to the outside world. The baby's head is then said to be 'engaged'.

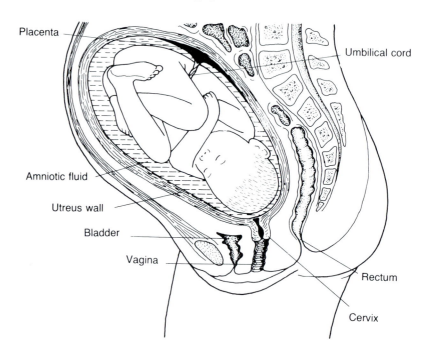

Fig. 4.2.19   The normal position of the baby at the time of birth

As **labour** begins, the cervix of the womb starts to open. The plug of mucus that has been filling the passage through the cervix during gestation slips out. A very slight amount of blood is usually passed out, but only enough to stain the mucus pink. The 'breaking of the waters' is an indication that labour is starting. 'The waters' is a general term used to describe the amniotic fluid which surrounded the baby in the womb. Although there are often several litres of this fluid, very little usually escapes when the membranes surrounding it break. This is because the baby's head traps a very small amount of liquid in front of it. It is only this which escapes. The much larger volume of liquid comes out with a rush when the baby is born.

The onset of muscular **contractions** of the womb is felt as a regular series of 'hardening' and 'tightening' sensations. The muscles in the wall of the uterus are beginning to shorten and become tight, thus pressing the baby downwards towards the vagina. With the first contractions, there is some discomfort but little or no pain. True labour pains are moderately strong and last for less than a minute at a time, but return with regularity every 15 to 20 minutes to start with.

Childbirth can be divided into three stages (Fig. 4.2.20). The first stage begins at the onset of labour and lasts until the cervix has widened to its maximum. The baby's head can now pass through it and into the vagina. The first stage is much the longest part of labour, but its actual duration can vary. For someone having her first baby, it is usually of the order of 10 to 14 hours. When the mother has been in the first stage for some considerable time, the contractions become more powerful. By now, they are usually only about three minutes apart, and they usually last for a minute or so. The mother may be given an anaesthetic gas to inhale if the pain is severe.

The second stage in a woman having her first baby lasts about an hour; in others it may be much shorter. By a combination of pushing and contraction of the uterus, the baby's head is slowly pushed down through the vagina until it becomes visible at the opening. A few contractions later, the widest part of the child's head moves through the ring of muscle that surrounds the vaginal opening. When the next contraction comes, the rest of the baby's head is forced out, followed by the rest of the baby's body.

Once the baby has been born, the second stage of labour is complete and the third stage begins. This is the stage in which the mother has to expel the placenta, or afterbirth. The placenta is the organ through which the baby has, until now, been receiving all its food and oxygen. It is close to the wall of the womb, while the cord runs through the vagina and is still connected to the baby.

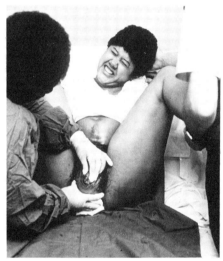 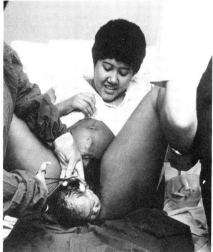

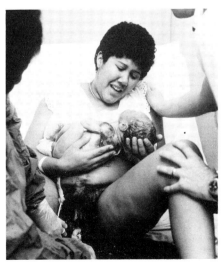

Fig. 4.2.20 Stages in childbirth

Two firm ligatures are tied around the cord a few centimetres away from the baby, and then a pair of sterile scissors is used to cut through the cord between the ligatures. Otherwise there would be considerable loss of blood when the cord is cut and the baby might bleed to death. Other mammals are able to bite through the umbilical cord without causing harm to the young. This method must have been used by prehistoric women. The mother experiences a return of mild contractions as the placenta is now forced out.

If things go wrong

Most mothers of large families will have experienced at least one birth which was complicated in some way.

Breech deliveries are quite common. In this type of birth, the baby comes 'tail first' instead of head first (see Fig. 4.2.21). Usually a midwife will have tried to correct the problem by turning the baby before labour begins. Although there is a slightly increased risk to the baby in a breech birth, the vast majority of babies delivered in this way with medical care come through none the worse for their experience.

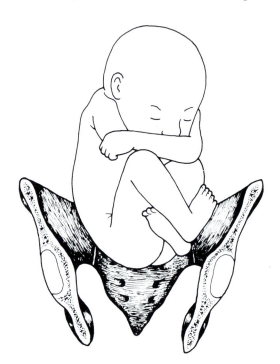

Fig. 4.2.21   Breech position of the foetus

Caesarian section is very frequently performed. There are various reasons why this procedure may be necessary. The most common is that the baby starts making it clear from an increase in heart rate that all is not well within the womb. The mother is taken to an operating theatre and given an anaesthetic. As soon as she is asleep, the surgeon makes a cut in the lower part of her abdomen, and delivers the baby through this opening. The incision is then carefully stitched up again. This method of delivery is now employed much more frequently than it used to be and has saved the lives of millions of babies throughout the world.

Growth

Growth of the human body begins at fertilization. It continues throughout the nine months in the womb and goes on until the end of adolescence. After that time, the skeleton has still about two per cent more to grow. All parts of the body change until the end of life.

The fertilized egg divides 44 times between fertilization and birth. From then until adulthood only four more divisions occur. The growth rate starts to get slower from about the fourth month in the womb and, apart from a spurt in adolescence, it gets slower and slower until it eventually stops.

Growth curves can be expressed in two ways. One way is to plot a person's height against age. This produces a gradually rising curve, with some uneven parts. The other is to plot the increase in height from one age to the next, expressed as a rate of growth per year. In a curve that shows rate of growth it is noticeable that from birth to about four years the rate of height decreases rapidly and becomes less steady until adolescence. Then there is a rapid increase in height gain until the peak of the adolescent growth spurt, when the height is increasing as fast as it was at the age of about two. After this point growth gradually slows down until it stops at about the age of 18. See Fig. 4.2.22.

Almost all organs show curves of growth similar to that of height: what differs is the timing of the curves. The brain, for example, is almost fully grown by the age of four. The reproductive organs have a slow start followed by a large increase at adolescence. These two organs show the extremes of development. Most parts of the body grow at an intermediate rate.

Growth of the skeleton is 98 per cent complete at the end of adolescence but fat continues to increase during the middle years of life. This ceases later and there is a gradual loss of fat in old age. Children enter the period of adolescence at very different ages. One child of 14 may be completely physically developed whilst another may not yet have started adolescence. The same is true, though perhaps less obviously, at all ages.

Some children grow rapidly from birth and stop growing early, whilst others take much longer to complete the growth process. Similar variations may occur in intellectual growth.

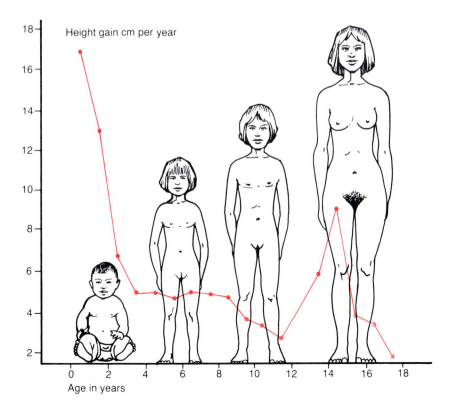

Fig. 4.2.22 Growth expressed as height gain per year

A person is able to grow until his or her bones are fully mature. There is a marked sex difference in the rate at which the skeletons of boys and girls mature. The average girl passes through the various stages of bone development one or two years ahead of the average boy. Girls grow less in adolescence than boys, so the average woman is shorter than the average man by about 10cm.

There are many factors affecting growth. Some of the most important are heredity, nutrition, season, and the state of health or disease. Tall parents tend to have tall children and these children are taller at all stages of growth. Genetic control operates not only in height but also, for example, in growth of the skeleton. There is a fairly high similarity between the age at which menstruation (menarche) begins in mothers and their daughters, in sisters, and particularly in identical twins. Parent–child similarities are much closer for height than they are for weight, which illustrates the importance of environmental factors on the latter. It seems likely that different sets of genes control growth at different times. Although genetic influences can also be detected in the times of eruption of the teeth and in the growth of other organs, it would be wrong to think that all growth is controlled by genes. It is possible, for example, to increase a child's bone age by overfeeding him from infancy.

Children, like flowers, grow in the spring, although they mostly put on weight in winter. The average height increase in spring is about twice that in winter. It appears that the length of daylight plays an important role in these seasonal variations in growth rate. The seasonal effect on weight gain, however, is probably due to changes in patterns of activity and of feeding.

Malnutrition in childhood delays growth, and malnutrition in the years before adolescence delays the onset of sexual maturity. Children subjected to temporary starvation recover completely provided it is not too severe and does not last very long. But malnutrition in the early months of life before birth may have permanent effects on growth. Children whose birthweights are very low may never be able to make up the deficit. Children who are overfed in the early months of life, on the other hand, may remain permanently heavier than other children. Their chance of suffering obesity in later childhood and in adult life is high. Such children are also taller than average at this early stage. The effects of overnutrition and malnutrition in later childhood are reversible. Adequate nutrition is necessary for normal growth but overnutrition is not a way of improving it. There are severe disadvantages which obesity causes.

Any illness in childhood causes a temporary interruption of growth. However, the regulatory processes are so finely adjusted that on recovery an immediate 'catch-up' period follows. The child then continues to grow at the normal rate. In short illnesses, such as measles, the interruption in growth is usually so small that it is not seen at all, but a prolonged illness of any kind slows down the rate of growth.

The long-term effects of illness depend on what it does to skeletal maturation. If the bones are retarded to the same degree as growth in height, then the final height after recovery can catch up and is likely to be normal. If skeletal maturation continues when growth in height has stopped, the final height is smaller than it would have been otherwise.

The control of growth   The control of growth is greatly influenced by hormones, the substances released from glands in one part of the body to exert an effect in other parts. Most hormones are under the control of the pituitary gland (see p. 190). Part of this gland releases **growth hormone** (see Fig. 4.2.23).

Fig. 4.2.23 The hormonal effects on growth

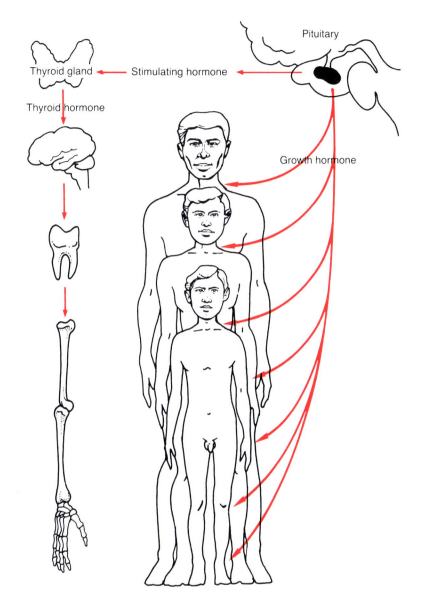

A child who fails to grow for lack of this hormone will show a marked increase in growth rate when given drugs containing it. On the other hand, giving growth hormone to a child who does not lack it, but is perhaps undersized for some other reason, will not help him or her to get any taller.

The reproductive glands are of special importance in the growth that occurs at the time of **puberty** (sexual maturity). The first noticeable event is an increased secretion of the hormones that stimulate the reproductive glands. These hormones are called **gonadotrophins**. Throughout childhood very small amounts are produced but as puberty approaches the amounts increase rapidly. This results in growth of the testes in boys and of the ovaries in girls, and the production from these glands of sex hormones. The sex hormones cause growth of the sexual organs and the breasts. Other hormones are responsible for the appearance of pubic and underarm hair.

At the same time special cells develop in the testes and the ovaries become able to release the eggs which have been present since before birth. See Fig. 4.2.24.

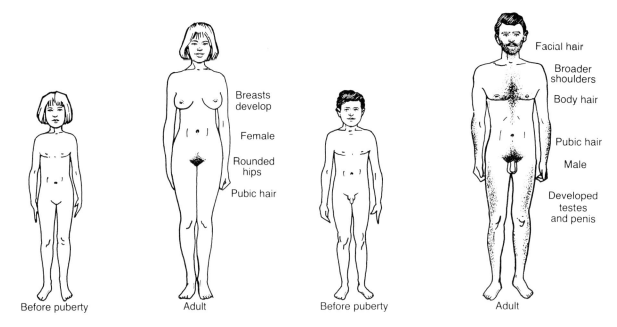

Fig. 4.2.24 Stages in growth

The growth spurt

Growth of the testes in boys usually starts at any age between 10 and 13, and continues for four or five years. At the same time pubic hair is also developing, but growth in the penis follows somewhat later. The peak of the adolescent growth spurt is not reached until a boy is fairly well into puberty, and facial hair appears only when this height spurt is beginning to stop.

Girls enter puberty earlier and may start to develop breasts between the ages of 8 and 13. The peak of the rate of growth in height also occurs earlier in girls than in boys (see Fig. 4.2.25). Growth of the uterus and vagina occurs at about the same time as growth of the breasts. Menstruation nearly always begins half way through the decrease in the rate of growth in height. On average, girls grow by only about another four centimetres after menstruation begins.

Fig. 4.2.25 Graphs relating to puberty in boys and girls

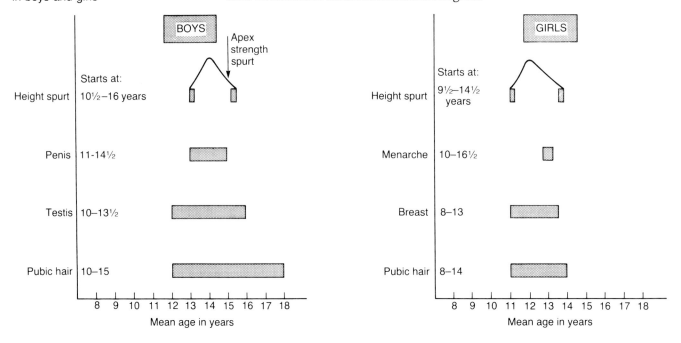

Almost every organ in the body takes part in the adolescent growth spurt, though not all to an equal amount. The brain is, by adolescence, almost its adult size. Most of the spurt in height is due to increase of growth of the trunk, rather than increase in the legs, but the spurt occurs in the legs before it occurs in the body breadth and trunk. The hands and feet mature first, then the legs, then the trunk. Boys gain in muscle bulk, and also in strength, whilst they tend to lose body fat. Girls, on the other hand, gain fat during adolescence and have a much smaller spurt in muscle bulk and strength. These changes are responsible for the obvious differences in appearance between the sexes but there are also differences in skeletal growth. A boy gets wider shoulders and a girl, wider hips. Adolescence may therefore be said to put the finishing touches to a child's growth. The last bones to complete their growth usually do so between the ages of 18 and 23 in boys. In boys 95 per cent have stopped growing by the age of 20 and 95 per cent of girls by the age of 19.

From early in foetal life the brain is nearer its adult proportion in terms of weight than any other organ of the body. Cell division in the brain occurs before birth and for the first two or three years after birth. During these early stages of development the brain is very sensitive to any sort of mishap. This explains the devastating effects of German measles, certain drugs, malnutrition and a decreased oxygen supply, on the foetus at this stage.

Ageing
(see Fig. 4.2.26)
There are physical and mental changes in the human body which are linked to old age. There is a progressive loss of natural hair colour towards grey, and the steady thinning of scalp hair which is more marked in men than in women. The end of egg production, with a drop in oestrogen, occurs at the menopause around fifty years old, accompanied by

Fig. 4.2.26 Elderly people

wrinkling of the skin and drooping breasts in older women. Both sexes suffer a gradual loss of elastic tissue in the supporting layers of the skin, which leads to wrinkling of the face, neck and other parts of the body. Breakdown of the cartilage lining and lubrication of such weight-bearing joints as the knees, hips and shoulders, causes discomfort and stiffness in limb movements.

A tendency to stoop results from loss of protein and calcium in the lumbar vertebrae, which become thin and compressed. An excess of abdominal fat and a decreased muscle tone tend to produce 'middle-aged spread' or paunchiness. The familiar old person's voice is related to changes in the muscles and cartilage in the larynx. Loss of elasticity in the lens and muscles of the eyeball often results in the inability to see print clearly.

The senses connected with feeling pain and vibrations, and recognizing objects by touch and odour, may become less efficient in later years. Hearing may deteriorate. This condition can be eased by suitable hearing aids. Loss of teeth has little to do with old age. It is more due to poor dental hygiene, lack of fluoride in water, and a high proportion of sugar in the diet.

There is a decrease in the ability to adjust to changes in outside temperature. Cold is poorly tolerated because the skin thickness is less. Shivering and other temperature controls do not function so well. Heart muscle activity may remain quite regular though less forceful and with less reserve for exercise. Shortness of breath may be present due to changes in lung function through loss of elastic recoil (**emphysema**) and lowered ability for the exchange of oxygen and carbon dioxide.

Mental changes due to thinning, softening and alteration of the brain tissue can occur with age. Memory, intellect and emotional feeling may all be altered to some extent. Powers of recall for recent events suffer some degree of degeneration. The alert older person is often aware of these failing qualities and may become depressed or gloomy.

A major change is that of hardening of the muscle walls of the arteries, called **arteriosclerosis**, or sometimes **atherosclerosis**. The arteries become narrowed and are lined with a material which reduces the blood supply to tissues. Since the arteries bring both oxygen and glucose for nutrition and energy release, the deprived tissues decay more rapidly than in the normal ageing process. This affects the brain cortex and heart muscles and especially the most important limb vessels.

Population growth and control

A population grows as a result of the excess of births and immigration over deaths and emigration. The rapid decline in death rates since 1945 in the developing countries was not accompanied by any fall in birth rates. This resulted in the biggest 'population explosion' in human history. If the rate of increase is maintained, the world's population will double itself during the rest of the twentieth century (see Fig. 4.2.27).

The populations of the developing countries grew by 45 per cent between 1950 and 1970. Those of the more developed countries increased by less than 25 per cent over the same period. The poorer countries already make up 70 per cent of the world's population. These developing countries may grow progressively poorer, because today their food supplies are barely sufficient. They cannot afford the extra demands on their resources brought about by uncontrolled population growth. The pressure of numbers on the limited medical services may destroy many improvements in health over recent years.

It is unrealistic to suggest that death rates should be allowed to rise again as a restraint on population growth. Thus the only feasible means

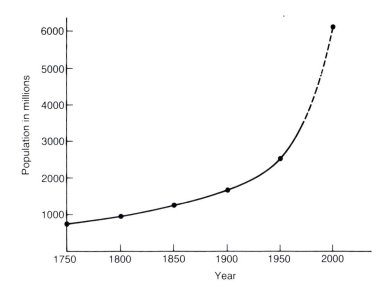

Fig. 4.2.27 The increase in world population

of limiting population is by birth control, and this is a major problem facing developing countries today.

Effective birth control cannot be achieved in a short period. Educational programmes are necessary before cultural resistance and ignorance can be overcome. Also, 45 per cent of the population in developing countries is under the age of 15. When these vast numbers of children enter the reproductive period of life, they will add further pressure to the growing population.

If birth control is reasonably effective the population of developing countries may grow to around four billion by the year 2000. This is an increase of 60 per cent, but if the programmes have no success the population may grow to more than six billion, which is an increase of 140 per cent. So, although population growth cannot be immediately halted, it can be checked and kept in step with development, but if it is unrestrained the consequences could be disastrous.

Methods of birth control
(see Fig. 4.2.28)

In the world as a whole, **male withdrawal** remains the most common method of contraception or birth control. It involves withdrawing the penis from the vagina just before sperms are released during ejaculation.

Douching or washing the inside of the vagina with water etc., following intercourse, is not an effective method. The sperms pass through the neck of the womb very rapidly, and those responsible for fertilization are probably the first through.

The **condom** (**sheath**) is by far the most widespread mechanical form of contraception in use. It became available as long ago as 1844. **Spermicides** (chemicals that kill the sperm) in the form of creams, foams, aerosols and C-films (water-soluble and plastic-coated with a spermicide) can also be used without medical supervision or advice. In most trials, they have been found to be less effective than condoms.

The effectiveness of a spermicide can be increased by some type of mechanical barrier placed in the vagina. Towards the end of the nineteenth century, well-designed rubber **diaphragms** and **caps** were developed. They became the main method suggested by the early birth control clinics. Used properly, these methods are more effective than the condom and are without side effects.

The **rhythm method** depends on the egg being released fourteen days before the next menstrual period occurs and on the fact that both the egg and the sperm have a short life. Fertilization is most likely at the time of ovulation. People practising the rhythm method are required not to have sexual intercourse for several days about this time. It is of decreasing importance in the Western World.

Intrauterine devices (**IUDs**) and **oral contraceptives** have come to be widely used since the late 1950s. Like many medical inventions the IUD was used before its action was understood. Usually, the device takes the form of a simple loop or spiral made of a material which causes no inflammation when it is placed in the uterus. IUDs appear to work by causing the migration of white blood cells from the bloodstream into the uterus. Here the cells probably destroy the egg between the time of fertilization and implantation. They are effective, do not need repeated attention once inserted, but can give rise to heavy and prolonged menstrual bleeding. The **contraceptive pill** (oral contraception) is a combination of ovarian hormones. These hormones prevent the release of the egg in the same way as ovulation is stopped during pregnancy.

Fig. 4.2.28 A selection of contraceptives: (a) the Pill, (b) condom and spermicide; (c) IUD; (d) cap and spermicide

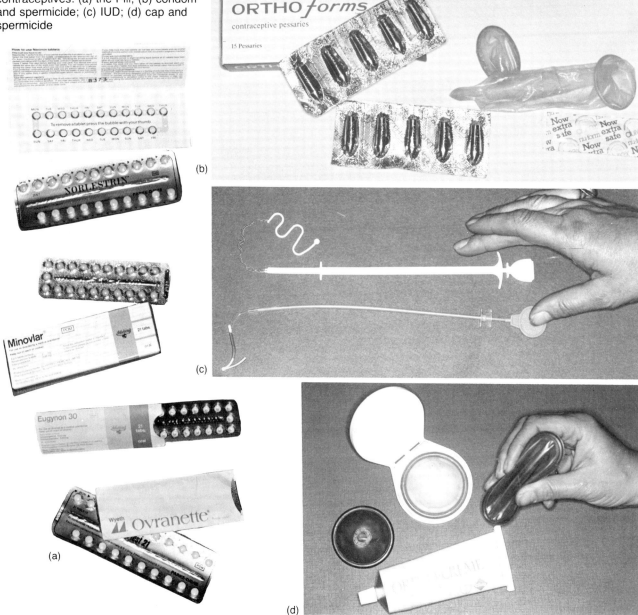

Since their introduction, the number of users of 'the pill' has reached an estimated thirty million throughout the world. As experience with the pill has grown, a number of side effects have been observed. Some, such as nausea (a feeling of sickness) are not important, but others, such as blood clotting may be very serious. The pill is the most effective reversible method of contraception, and risks of death due to unwanted pregnancies equal or outweigh the risk of death due to thrombosis (blood clotting).

The pill is more effective than IUDs; IUDs are more effective than caps; caps are more effective than condoms, spermicides, withdrawal or rhythm methods, and douching appears to be the least effective of all.

All the methods so far mentioned have been reversible, that is, a person can still have a family when the chosen contraceptive method is stopped. However, there is the irreversible method of surgical control. Male and female sterilization is becoming increasingly widely used. In female sterilization the fallopian tubes must be closed and an abdominal operation is necessary. On a world scale the use of this method is limited by the available medical resources. Sterilization of the man by cutting the vas deferens (performing a **vasectomy**) is a simple operation by comparison and can be done in only five to ten minutes.

Neither operation interferes with the sexual activity of the person concerned. The ovaries and testes continue to make sex hormones which pass into the bloodstream and control sexual behaviour. The blocked eggs and sperms are reabsorbed by normal bodily processes.

Disease and the reproductive system

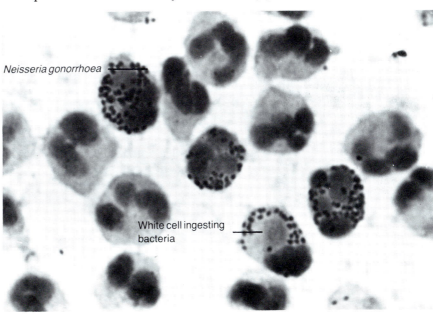

Neisseria gonorrhoea

White cell ingesting bacteria

Fig. 4.2.29 *Neisseria gonorrhoeae –* the causative agent in gonorrhoea

Gonorrhoea

The causative organism of **gonorrhoea** was discovered by Neisser in 1879. It is now called *Neisseria gonorrhoeae* (Fig. 4.2.29) and is rapidly destroyed by most antiseptics even in weak concentrations. Its survival outside the human body is limited and it is particularly sensitive to lack of moisture. This factor reduces the spread of the disease from contaminated towels, bed linen and toilet seats.

Transmission of the disease invariably occurs by sexual contact with an infected person. Very occasionally infection may be acquired from infected towels etc. In the newborn, the eye infection, conjunctivitis, may develop because infective material from the birth canal can get into the infant's eyes.

The parasite normally enters the body via the urethra or vagina. It causes the production of a thick discharge from these sites. The infection may travel to the prostate gland in men or the fallopian tubes in women. Spread of the infection via the patient's bloodstream may also take place.

The incubation period of gonorrhoea may be as long as 14 days. In men the earliest symptom is pain on passing urine. Later there is a discharge, an abscess may develop, and the infection may spread through the reproductive organs.

In women, the urethra and the neck of the womb may both be infected. Pain on passing urine is a common symptom. Vaginal discharge is slightly less common. Infection of the rectum is very frequent. In approximately 50 per cent of all cases of female gonorrhoea there are no symptoms at all. These carriers are more likely to transmit the disease than are patients who show symptoms.

In both sexes tests are necessary to find out if the organism is present in the reproductive organs. Specimens of the discharge from men and women are examined on culture dishes in the laboratory.

Until the introduction of **sulphonamide** drugs in 1935, the treatment of gonorrhoea relied entirely on the use of antiseptics. The antiseptic varied but the most common was potassium permanganate. The introduction of sulphonamide-related drugs revolutionized the treatment of gonorrhoea. A week's treatment with sulphapyridine cured 90 per cent of patients.

Unfortunately, by 1943, because of the development of drug resistance, the cure rate had fallen to about 50 per cent. Help came with the introduction of penicillin in the same year. Four or five injections over 10 to 12 hours were necessary but the rate of cure was almost 100 per cent. By 1947, long-acting penicillin preparations were available which gave equally successful results with a single injection.

Reports of penicillin failures began to appear in 1958, and laboratory investigation showed that penicillin-resistant strains were responsible. By 1971, the percentage of resistant strain was 80 per cent in some parts of the world. Resistance to other antibiotics such as streptomycin and the tetracyclines has developed since 1960. Despite claims for newer antibiotics and renewed interest in the sulphonamide drugs, penicillin remains the most widely used antibiotic in the treatment of gonorrhoea.

Following treatment, patients are advised to return to the doctor at intervals over a period of three months. In men, successful treatment is indicated by disappearance of the discharge and clear urine which is free from pus. In women, repeated tests are necessary before a cure can be confirmed. In every case of gonorrhoea the patient is asked who was the source of infection, and he or she should also be questioned about further contacts so that they can be offered help. Gonorrhoea is highly infectious in the incubation period even before there is any noticeable discharge.

Syphilis

This disease is caused by a type of bacterium called *Treponoma pallidum* (Fig. 4.2.30). A sore called a chancre appears where the infection has taken place. After three to six weeks a body rash, sore throat and fever usually occur. Loss of hair may take place after four to eight weeks. Ulcers develop near the mouth and, in the female, near the entrance to the vagina. If the disease remains untreated, it may attack the heart and nervous system, leading to blindness and brain damage.

If a pregnant woman is infected with the disease, the developing foetus can contact it via the placenta. Sexual contact, during intercourse, is the usual method of spreading the disease but the bacterium can also be transmitted via skin wounds and tatooing with contaminated needles.

The antibiotics penicillin and erythromycin are used to treat the disease.

Fig. 4.2.30 *Treponoma pallidum –*
the causative agent in syphilis

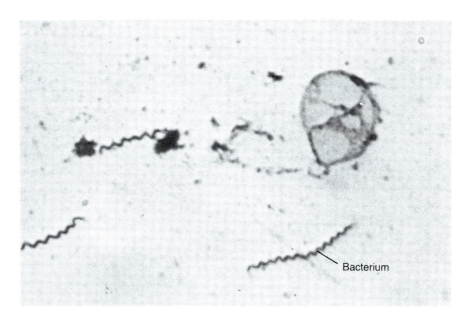

Bacterium

Chlamydia This is the most common sexually transmitted disease in Britain. The organism *Chlamydia trachmatis* (see Fig. 4.2.31) is responsible for at least half of all non-specific vaginal and urethral infections reported to clinics each year – about 75 000 cases.

It was first thought to be a virus, because it can only live inside another cell. It is, however, a bacterium. In men it leads to itching and a burning sensation in the urethra. In women it largely fails to show symptoms, although it can lead to a discharge from the vagina.

If left untreated, the infection may remain dormant for a time before spreading into the uterus and fallopian tubes. This may lead to infertility. The infection may also lead to conjunctivitis, an infection of the conjunctiva of the eye. Newborn babies can contract this eye infection while being born through the vagina of an infected woman. This can be particularly serious because if it is not treated the infection can spread to the lungs, causing pneumonia.

If it is recognized in its early stages, *Chlamydia* can be easily treated with a two to three week course of the antibiotics tetracycline or erythromycin.

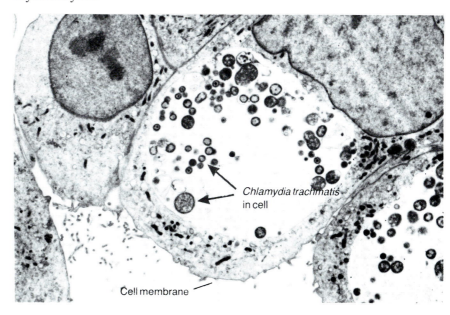

Chlamydia trachmatis
in cell

Cell membrane

Fig. 4.2.31 *Chlamydia trachmatis*

Questions: Domains I and II

1 The graph shows part of the menstrual cycle and hormonal control.

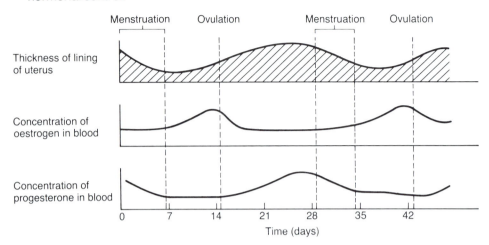

a What do you understand by (i) menstruation (ii) ovulation? (I)

b How long is the menstrual cycle shown in the graph? (II)

c Which hormone is at its highest level
(i) just before ovulation;
(ii) just before menstruation? (II)

2 The table shows the use of the different types of contraception as a percentage of the total use.

Method of contraception	%
Pill	36
Cap	3
Intrauterine device (IUD)	15
Sheath	25
Withdrawal	5
Rhythm (safe period)	1
Male vasectomy	3
Female sterilization	4
Other	8

a Plot the percentages in the form of a histogram. (II)

b Which method is (i) most commonly used; (ii) least commonly used? (II)

c Which method of contraception is most likely to lead to an unwanted pregnancy? Give a reason for your answer. (I)

3 The diagram shows the blood flow between the embryo and the placenta.

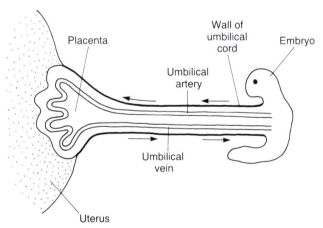

a Which blood vessel shown will have the greater concentration of oxygen? (I)

b Name one substance that will be in greater concentration in the umbilical artery than in the umbilical vein. (I)

c Explain how the blood is pumped along the umbilical artery. (I)

d Simple sugars pass along the umbilical vein to the embryo. Give one use that the embryo makes of the sugars. (I)

e What is the advantage of the wall of the placenta being folded? (I)

4 The table below shows the number of twin births per 1000 pregnancies for mothers of different ages.

Age of mother in years	15	20	25	30	36	40	45
Number of non-identical twins/1000 births	0.8	4	7.2	10.4	15.6	12	4
Number of identical twins/1000 births	2.8	3	3.2	3.6	3.8	4	4

a Plot these results on graph paper. (II)
b State two conclusions that can be made from the graphs. (II)
c Explain how identical twins are produced. (I)
d Explain which type of twins may be of different sexes. (I)

5 The diagram shows a human gamete.

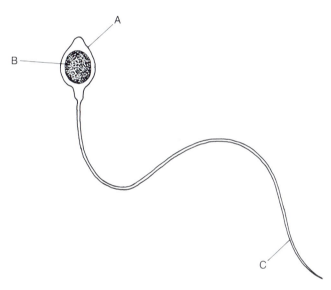

a Name the structures labelled A–C.
b State the name of the part of the gamete where chromosomes are found.
c Give one function of the part labelled C.
d During reproduction, zygotes are produced by the fusion of gametes. What is the name given to this process of fusion?
e What is the name given to the interval of time between the fusion of gametes and the birth of the baby?
f State one difference between the structure of the nucleus of a gamete and that of a normal body cell.
g Which chromosome determines the sex of a female baby?
h Give two reasons why it is beneficial to breast feed a baby.
i State two reasons why it is considered advisable that girls between the ages of 10 and 13 years should be vaccinated against German measles (Rubella).
j Briefly explain how it is dangerous to the unborn baby if its mother suffers from anaemia.
k Which mineral would be given to help correct the anaemic condition? (I)

6 The graph shows births and deaths in the United Kingdom.

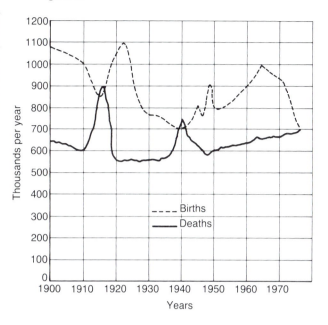

a In 1900, what was the
(i) birth rate; (ii) death rate? (II)
b In 1976, what was the
(i) birth rate; (ii) death rate? (II)
c In which year did the birth rate reach a maximum? (II)
d In which two years did the death rate reach peaks? (II)
e In 1965 was there an increase or decrease in population? Give a reason for your answer. (I)
f In the 1970s the birth rate was falling. Suggest two effects this would have on community services. (I)
g In developing countries there are now less infant deaths than there were fifty years ago. Give three reasons why there has been a change. (I)
h One way of controlling the birth rate is by contraception. What is meant by contraception? Name two methods of contraception. (I)

7 The table shows the increase in the number of people suffering from sexually transmitted diseases.

Year	Number of males and females in the UK that were new patients for treatment
1961	68 068
1966	82 979
1971	139 472
1978	168 804

a Suggest two reasons why the numbers have increased over 17 years. (I)
b Between which years was the greatest number of new cases reported, and how many? (II)
c State two precautions which can be taken to prevent further increases in sexually transmitted diseases. (I)
d What symptoms would be noticed in the male and female suffering from gonorrhoea? (I)
e How could gonorrhoea be treated? (I)

8 The diagram shows the female reproductive organs.

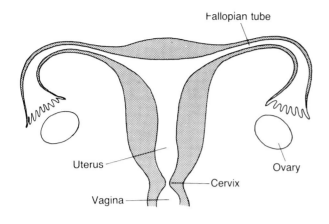

a Explain why a woman with blocked Fallopian tubes cannot become pregnant in the normal way.

b Use the following account and the diagram above to answer the questions.

A woman with blocked Fallopian tubes can have a 'test-tube' baby. She is given hormones to increase the number of eggs maturing in her ovaries. A doctor, using a fine tube through the body wall, searches for and collects several eggs from the surface of the ovary just before they are released naturally. The eggs are then put in a culture solution in a culture dish. Semen containing active sperms is added and fertilization usually occurs. Three days after fertilization, embryos of between eight and sixteen cells have formed. Two or three of these embryos are gently transferred by a fine tube via the cervix into the uterus. If the process is successful, at least one of the embryos develops into a baby.

 i Suggest why the doctor wants to increase the number of eggs maturing in each ovary. (I)
 ii Suggest why the doctor wants to collect eggs just before they are released naturally and at what stage of the menstrual cycle the doctor would do this. (I)
 iii Suggest features of the culture solution which are essential for a successful test-tube baby. (I)
 iv Suggest why the doctor waits three days before transferring the embryos to the uterus. (I)
 v Suggest why, when the embryos are put in the uterus, the tube is passed through the cervix and not through the body wall. (I)

c Explain why the term 'test-tube baby' may not be the most accurate one to use for a description of this process. (I)

d During pregnancy, a sample of amniotic fluid can be withdrawn from the womb with a fine syringe. When examined with a microscope, the fluid often contains cells of the embryo.
Explain how these cells can be used to determine
 i the sex of the embryo;
 ii Down's syndrome. (I)

e During pregnancy a mother's total body mass may increase by between 9.5 kg and 13.0 kg. The bar chart shows the components which cause this increase.

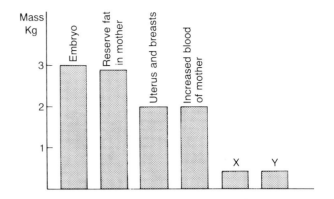

 i Suggest a name for the components represented by X and Y on the bar chart. (I)
 ii State one reason for the increased blood mass of the mother. (I)

f State the changes which normally occur to each of the following during the birth of a baby:
 i muscles of the uterus wall;
 ii muscles of the cervix. (I)

9 The diagram represents the life cycle of a human.

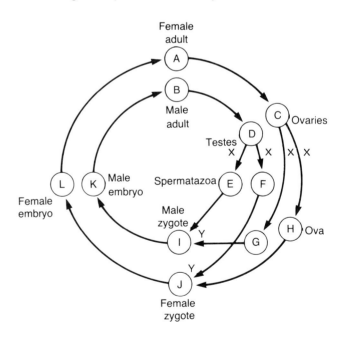

a What is the chromosome make-up (XX, XY, X or Y) of the different nuclei represented by the circles A–L? (I)

b Name the process that occurs at X when spermatozoa and ova are formed. (I)

c Name a process that occurs at Y when spermatozoa and ova form zygotes. (I)

10 Many parts of the female reproductive tract are damaging to sperms. The pH of vaginal secretions is 3.5–4.2. A normal volume of semen discharged during intercourse is 3.4cm^3. This is important to neutralize the acid vaginal secretions. Too great a volume is not helpful for fertilization. It has been found that the total amount of carbon dioxide given out by a sperm is the same whether it has had a short and active or long and sluggish existence. Carbon dioxide makes sperms inactive. When the sperms are deposited in the female reproductive system, they lose their ability to fertilize an egg within forty-eight hours at the most.

a List the factors which affect the survival of sperms. (I)

b Explain how the factors you have listed affect the sperms. (I)

11 The graph below shows average measurements of a foetus from the third to the tenth lunar month of pregnancy (1 lunar month = 28 days). Study the graph and answer the questions that follow.

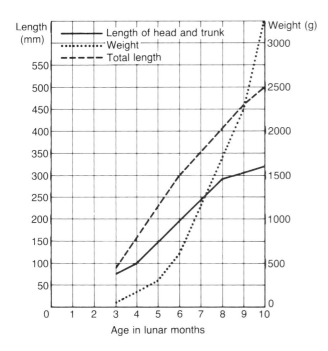

Age in lunar months

a When was the period of greatest rate of increase in weight? (II)

b How long are the legs of the foetus at the end of the 10th lunar month? (II)

c Which of the measurements was the first to slow its rate? (II)

d At six months, what is the ratio of the total foetal length to the length of the head and trunk? (II)

e Determine the period when the greatest rate of brain development probably occurred. (II)

f Suggest the advantage to the mother of the slower rate of growth of the head and trunk between the 8th and 10th lunar months. (I)

Answers to Questions: Domains I and II

Chapter 1.1 (p. 19)

1 a **A** has long fingers with tips modified for grasping branches of trees in which it lives.
   B has long fingers, not as long as **A**, suggesting that part of its time is spent in trees.
   C has a shorter, more powerful looking hand. Probably not for life in trees.
 b The opposable thumb and fingers indicates the ability to form a precision grip as well as the ability to grasp objects.
 c The foot of **H** is straighter, with all toes directed forward. The big toe is very close to the other toes. The foot is not capable of grasping things.
 d Foot **H** is best suited for walking erect as it appears to be best suited for balance. All the toes touch the floor together with the heel and ball of the foot.
2 a **A** has larger cranium, smaller canines, smaller ridges over the eyes, smaller jaw and a smoother, non-crested top to the cranium.
 b Man has a larger brain and therefore is more intelligent. Man's teeth are less well adapted for a carnivorous diet. Man's jaws are not as strong as those of the gorilla.

Chapter 1.2 (p. 24)

1 a See graph.

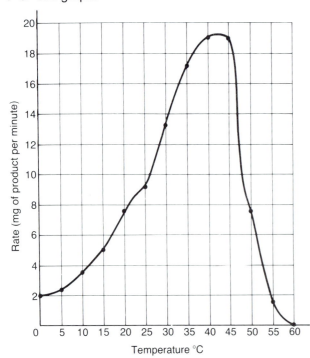

Temperature °C

c At 17°C the energy supplied to the reaction by the heat is high enough to increase the collision between enzyme and substrate molecules in the reaction. Therefore a lot of product is being formed.
 d Above 45°C, the enzyme becomes denatured. This means that it can no longer form the product from the substrate as the high temperature has altered its molecules.
 e Acidity/alkalinity, concentration of substrate, concentration of enzyme.

Chapter 2.1 (p. 31)

1 a Sunlight.
 b If the mice were eliminated, the rabbits would not have so much competition for food and their population would increase. *Or* if the mice were eliminated, the owls would eat more rabbits and the rabbit population could decrease.
 c Humus → Earthworm → Domestic fowl → Man.
 d Peas.

2 a Photosynthesis is taking place in the leaves. Therefore the leaves are using up carbon dioxide which is an acid gas. Therefore the contents of the tubes become less acid.
 b There is a greater amount of leaf tissue in D so more carbon dioxide is used.
 c E acts as a control to prove that it was the leaf tissue that was responsible for the change in the colour of the indicator.
 d Tube D would contain most oxygen because there is more tissue there to produce it during photosynthesis.
 e Tube E would contain most oxygen. In the other tubes, it would be used up during respiration.

Chapter 2.2 (p. 50)

1 a To demonstrate how effective a toothpaste is at preventing bacterial growth.
 b An antiseptic.
 c It does not cause disease.
 d Heat it in a strong flame.
 e 25°C. Bacteria should not be incubated at body temperature in school laboratories because accidentally introduced pathogens might grow at this temperature.
 f Toothpaste Y.

2 a 5 days.
 b Day 8.
 c 40.5°C.

d i They are reproducing quickly.
  ii They are dying.
e 37°C.
f Coughing and a rash.
g i Natural immunity means having naturally occurring antibodies to counteract a particular pathogen.
  ii Acquired immunity means that the body has to be encouraged to produce antibodies as a result of being vaccinated with either the dead or weakened pathogen or with serum from an infected animal.
h The environmental Health Officer.
i Typhoid, measles, poliomyelitis, smallpox, yellow fever, cholera, diphtheria, food poisoning, tetanus, tuberculosis, whooping cough, malaria.
j Water – cholera. Insects – malaria. Food – food poisoning.

3 a i A disease, regularly found, and possibly originating in a particular country.
  ii A single-celled animal.
  iii An organism which lives and feeds on another living organism of a different species.
  iv A carrier of disease.
  v A substance produced by a living organism which kills bacteria.
b The lymph glands produce many cells which ingest and kill the parasites. When the body is infected these glands make more cells than usual.
c Because the parasite affects the nervous system.
d The drug passed through the parasite too quickly.
e Serum albumin made the drug stay in the parasite.
f It means that a drug will only affect a particular target, in this case, the parasite.
g Poisonous arsenical compounds.
h Older drugs could be used on specific targets i.e. the parasite only.

4 a Passive.
b 40 days.
c Three days.
d Passive because there is almost immediate protection.
e There is a safe level with the passive method but not with the active method.

5 a The river water – probably from sewage.
b The corned beef became contaminated from untreated water used in its production.
c The microbes were transferred to the meat slicer from the corned beef.
d The incubation period.
e Tracing and isolation contacts. Vaccination of the community. Confiscation of contaminated food.
f i Large number of people within a community show symptoms.
  ii There is a rapid spread of the disease.
  iii New cases being reported each day over a given period of time depending on the nature of the disease.

Chapter 2.3 (p. 57)

1 a Lowers the pH.
b Bacteria eat the sugar and produce acid waste products.
c Two minutes.
d 28 minutes.
e Bacteria and food debris.
f Regular dental checks. Regular brushing. Reduction of sugar in the diet.
g Calcium. Phosphorus.

2 a A.
b Children had less tooth decay in A.
c (i) 75. (ii) 6.
d (i) A 76. (ii) B 179.
e 12.5%.
f Too much sugar in the diet. Poor dental hygiene. Lack of dental checks and treatment.

3 a Cotton.
b Nylon.
c Wool.
d Slower heat loss.
e A small volume of water loses heat more quickly than a larger volume.
f Temperature. Air movement.
g Use the same mass of fabrics.
h i Wash easily and dry quickly.
  ii Absorbs sweat easily. Can be boiled for cleaning.
  iii Sweat and dirt are breeding grounds for micro-organisms.
i i Prevents flies getting at refuse.
  ii Prevents bacterial growth.
  iii Prevents bacteria spreading into the toilet from drains because of the water seal.
  iv Slows down bacterial growth on meat.

Chapter 2.4 (p. 68)

1 **A** – The front bedroom. An inflammable nightdress is being worn close to the electric fire. A faulty electric blanket is being used. Cleaning fluid in an unmarked bottle has been left within reach of children.
  B – The bathroom. An electric lamp is being handled with the person in contact with water. A puddle makes the floor slippery.
  C – The centre bedroom. A slippery carpet is on the floor. Someone is smoking in bed.
  D – Back bedroom. A heating stove has fallen over. The dog is chewing the electric flex.
  E – The kitchen. The oven door allows the child to crawl in and tip over the oven. Pots on the stove have their handles in reach of children. Spillage can cause serious burns if hot water is carried from the stove instead of taking the teapot or cups to the stove. An unsafe platform is being

used to reach something stored high up. The electric socket is over-loaded and the trailing flex could trip someone.

F – On the stairs. Toys have been left on the stairs where they can cause tripping.

G – The living room. A worn out rug can trip someone over. The fire is left unguarded. Cigarette ends are left smouldering. A glass window overlooks an area where children play.

H – The garage. An amateur mechanic is trying to mend the brakes. Cleaning fluid is in an empty drinks bottle in reach of children.

I – The basement. A faulty stair has not been repaired. A gas leak is being searched for with a naked flame.

J – The laundry–playroom. A frayed electric flex trails through water. A badly designed and unsafe TV set is being used. A power tool is not being used safely.

K – In the garden. A tin of weed killer has been left on the lawn. It could be poisonous or could cause a fire. Unattended electric shears are within a child's reach. The ladder is unsafe.

a **A** remains fresh. Most bacteria were killed. Refrigeration slows growth.
B becomes sour. Most bacteria were killed. Those remaining multiply in the warm room.
C begins to go sour. Most bacteria were killed. There is a slower rate of growth than **B** because of the lower temperature.
D is slightly sour. Bacteria are not killed but refrigeration slows down the growth rate.
E is very sour. Bacteria are not killed and there is a high rate of growth in the warmth.
F is sour. Bacteria not killed but the cool temperature makes the growth rate slower than **E**.

b To make sure no bacteria were in the containers to cause souring.

c Pasteurization.

d i Canning – removes air and kills bacteria.
  ii Drying – removes water and prevents bacterial growth.
  iii Pickling – kills bacteria with acid.

Chapter 2.5 (p. 78)

1 a Natural lakes. Artificial lakes. Shallow wells. Deep wells. Rivers.
  b London.
  c Water from sewage effluent.
  d The lid prevents pollution from soil and vegetation and surface rain water.
  e It prevents water from surrounding soil percolating into the well.
  f Flocculants.
  g It has been filtered by a great depth of subsoil and rock.
  h Algae have to be removed. Mineral nutrients, e.g.

nitrates, have to be removed. Pollution from soil run-off will have to be prevented.
  i Bacteria may be pathogenic. Algae may be poisonous. Clay gives the water a 'dirty' appearance.
  j It protects the water from bacterial contamination during storage.

2 a (i) Process Y. (ii) Process X.
  b It loses pathogenic organisms from A–B. It loses all organic material capable of break down to methane gas.
  c Methane.
  d It can be used as a fuel for releasing energy as heat.
  e Fertilizer for agriculture.
  f Reduces chances of contamination with pathogenic bacteria. Dirty water would be useless for drinking and washing.
  g Bacteria. Suspended material. Algae.
  h (i) Chlorine kills bacteria. (ii) A covered reservoir prevents contamination by impurities in surface water.

3 a From bacteria on food with expired sell-by date. From the dog. From the organic debris on the floor. From poorly maintained toilet. From the open dust bin. From the inability to wash hands after using the toilet. From the dirty appearance of the shop assistant. From the money being handled near food.
  b Clean all surfaces thoroughly. Improve toilet facilities. Repair the wash basin. Dispose of all food with expired sell-by date. Insist on better hygiene standards from shop assistant. Improve facilities for refuse disposal, i.e. have a closed dust bin. Incinerate cartons etc.
  c *Cardboard packets* – collapse them and tie them in a bundle.
Bottles – keep them in a box.
Cans – keep them in a box unless they have been used as containers for food, in which case they should be kept in a sealed dust bin.
Food scraps – wrap them in newspaper and put them in sealable polythene bags.
  d This is dumped on a controlled rubbish dump. This means that it is covered with a layer of soil and regularly disinfected or sprayed with insecticide in an enclosed area.
  e An old refrigerator should have its door removed, then given (sold) to a scrap metal merchant. Garden refuse should be composted.

Chapter 2.6 (p. 88)

1 a 32 million tonnes per year.
  b Industry.

c

	Power stations	Vehicles
Nitrogen oxides (tonnes per year)	2	6
Hydrocarbons (tonnes per year)	2	12

d Petrol driven motor vehicles appear to produce most of the atmospheric pollution shown on the bar chart.

e Carbon monoxide is dangerous because it combines with haemoglobin in the blood, preventing oxygen being carried by the haemoglobin.

2 a

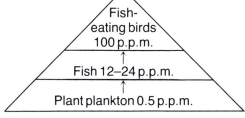

Fish-eating birds 100 p.p.m.

Fish 12–24 p.p.m.

Plant plankton 0.5 p.p.m.

b The concentration of mercury builds up through the food chain because it accumulates in the tissues of the organisms at each level.

c i Mercury reached the blood via the diet. They ate fish which contained mercury which became absorbed into the blood stream through the intestine.

ii The nervous tissue.

3 a Fish.

b It was used as an insecticide.

c i When DDT was sprayed on the land as an agricultural insecticide it remained in the soil until it was washed into rivers which took it into the sea.

ii The concentration of DDT builds up through the food chains in the sea because it accumulates in the tissues of the organisms at each feeding level.

d It has to come on to land in order to lay its eggs.

4 a Protein.

b Carbon monoxide, sulphur dioxide, particles (soot).

c Plants near the motorway have a higher concentration of nitrogen and therefore the nitrogen can be used for protein synthesis. The plants are therefore richer in protein for the animals to feed on. There are also fewer predators to feed on the animals on motorway verges.

d The salt solution would prevent the normal process of osmosis from occurring in the roots of the plant because there would be a strong solution outside the root compared with the inside. Water would tend to be drawn out of the plant by osmosis.

5 a Ants and wild thyme.

b When it feeds on the flowers and developing seeds of the wild thyme. Also when it feeds on nectar.

c When the larvae eat the ant grubs.

d Wild thyme.

6 a i They lay eggs in water.

ii Old farming methods relied on water being in the pond for the horses and for washing carts. As these methods have largely changed, the pond water is no longer necessary.

iii The fast flowing water of the brook would carry the frog spawn away.

b (i) Salt. (ii) For de-icing road surfaces in winter.

c The salt water draws water out of the frog through the selectively permeable skin by osmosis. In the same way food is preserved in salt.

d He has dug the pond deeper to retain more water and he has added spawn to the pond.

7 a i The radioactive materials become washed into the soil and are taken up in the roots of plants. Eventually they reach the stems and leaves.

ii The materials enter plants as in (i). Cows eat the affected plants and the radioactive chemicals become incorporated into milk when it is produced by the cows.

b There is the problem of disposing of the radioactive waste material. This is not the case where solar and tidal power are used.

c Iodine is incorporated into the molecules of the hormone, thyroxine. This is produced in the thyroid gland. If the iodine is radioactive it can be traced as an accumulation in the thyroid gland.

8 a The direct damage to wildlife by killing various animals on the road.

b The destruction of natural habitats by building roads.

c The pollution of the atmosphere by exhaust fumes.

Chapter 3.1 (p. 97)

1

	Skull	Hip	Finger	Vertebral column	Shoulder	Knee
Ball and socket		✓			✓	
Gliding			✓	✓		
Fused	✓					
Hinged						✓

2 a 1 Gliding   Fingers
    2 Pivot   Wrist (radius and ulna)
    3 Hinged   Elbow

b Striated/striped/voluntary/skeletal.

c One contracts while its partner relaxes (antagonistic pair).

3   Protection   Skull
      Support   Vertebral column
      Movement   Jointed limbs

4 a A: scapula (shoulder blade). B: Humerus; C: Ulna; D: Radius.
  b Tendon. It must be non-elastic because it attaches muscle to bone. When the muscle contracts it must transmit its pull through material that will not stretch.
  c Cartilage.
  d Biceps by contraction.
  e (i) pelvis (ii) synovial fluid (iii) antagonistic (iv) insertion (v) origin (vi) muscle tone.

5 i It would be too heavy if it were made of solid bone.
  ii It would not be as strong because of the 'girder' arrangement which resists compressive stress.

6 a See graph.
  b 30 minutes. (c) 10–20. (d)(i) 9.5; (ii) 4.5.

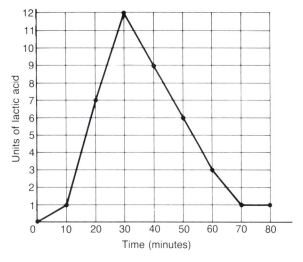

7 a Movement often needs to be quick to avoid harmful stimuli.
  b Smooth muscle is often responsible for slow reactions such as peristalsis. This allows food to remain in the gut long enough for digestion.
  c Ileum. (d) It must not suffer from fatigue. (e) The cells are branched and form a network.

Chapter 3.2 (p. 122)

1 a (i) male; (ii) 16–20; (iii) 13–15; (iv) 10g per day.
  b 400g.

2 (a) Apples; (b) 50g; (c) High temperatures destroy ascorbic acid; also if sodium hydrogen carbonate is used in the cooking process, the ascorbic acid will be neutralized.

3 a (i) Milk; (ii) 75%; (iii) Rice; (iv) Milk.
  b Beans; c Vitamins and minerals; d Vitamins are needed to help enzymes to work. Without them we suffer from deficiency diseases. Minerals are also often used to help enzymes work. We suffer from

deficiency diseases if we lack them, e.g. anaemia – lack of iron; e Rice.

Chapter 3.3 (p. 133)

1 a Gillian because the vegetables provided minerals and vitamins; the beef provided protein and fat; the potatoes provided carbohydrate; the yoghurt provided a small amount of all the classes. Anne's meal lacked vitamins, minerals and protein.
  b Milk/yoghurt/cheese.
  c All chemical reactions taking place in your body need water. Water is needed to dilute harmful waste products for excretion.
  d An energy store. Insulation.
  e In the liver.
  f The amino acid is split into the amino group (NH_2) and the acid group (COOH). The NH_2 group goes to the formation of urea (NH_2)CO.
  g Ileum.
  h

Region	Juice	Enzyme	Food acted upon	Product
mouth	saliva	amylase	starch	maltose
stomach	gastric	pepsin	proteins	peptones and proteoses
pancreas	pancreatic	lipase	fat	fatty acids and glycerol.

2 a A: stomach, B: pancreas, C: gall bladder, D: colon, E: rectum, F: appendix.
  b (i) 10%. (ii) There are 2% sugars in the oesophagus. This means that starch must have been digested to sugar by a carbohydrase. (iii) There is no protein digestion in the mouth. There are no free amino acids in the diet. (iv) 76%. (v) It has been diluted by stomach secretions. (vi) Acid secretion by the stomach. (vii) It is completely digested to sugar. (viii) I.

3 a A blood capillaries, B lacteal, C epithelium, D lymph vessel.
  b Ileum.
  c Blood capillaries.
  d Glycerol and fatty acids.
  e Lacteal.
  f Thin walled/large surface area/rich blood supply.

4 a (i) Hepatic artery; (ii) hepatic portal vein; (iii) hepatic vein.
  b The explanation could be related to the function of the pancreas as a digestive gland or as an endocrine gland. In the first instance removal would mean that digestion would be incomplete because of lack of pancreatic juice. In the second case the inability to produce insulin and control blood sugar levels would lead to death.

5 a See graph.
  b 4 minutes.

c The enzyme is inactive at 10°C and is denatured at 60°C.

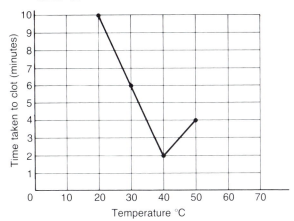

Chapter 3.4 (p. 158)

1 Number of white cells/mm^3 = 6000. Number of red cells/mm^3 = 5 000 000. Number of platelets/mm^3 = 300 000.
Function of plasma. Acts as a medium for the transport of products of digestion, vitamins, minerals, hormones, carbon dioxide (mainly as sodium hydrogen carbonate) and urea.
Function of white cells. To help in the body's defence against disease.
Function of red cells. For the transport of oxygen.
Function of platelets. To help in clotting.

2

Substance carried	From	To	How carried
Carbon dioxide	tissues	lungs	Sodium hydrogen carbonate.
Products of digestion	intestine	liver and then to tissues.	In solution in plasma or in suspension.
Urea	liver	kidneys	In solution in plasma.
Hormones	endocrine glands	tissues	In plasma.

3 Group A. Antibody b. Group B. Antibody a. Group AB. No antibodies. Group O. Antibodies a and b.

4 a (i) Right ventricle. (ii) Left ventricle. (iii) Lungs. (iv) Left atrium(auricle). (v) Ileum or liver. (vi) Right atrium.
  b Arteries carry blood from the heart, veins carry blood to the heart.
  c (i) Vena cava. (ii) Aorta.
  d Carbon dioxide.

5 a Hepatic vein, vena cava, right atrium, right ventricle, pulmonary artery, pulmonary vein, left atrium, left ventricle, aorta, renal artery.
  b i

Arteries	**Veins**
Very muscular walls	Not very muscular
Very elastic walls	Not very elastic
No valves	Valves present
Thick walls	Thin walls

  ii

Arteries	**Veins**
Carry blood away from the heart.	Carry blood towards the heart.
Except for the pulmonaries, they carry oxygenated blood.	Except for the pulmonaries, they carry deoxygenated blood.
Most carry blood which contains much digested food.	Most carry blood which contains much waste material.

6 a Because there is so much variation in individuals' pulse rates, an average would provide a more valid way of deducing the effect of exercise on pulse rate.
  b More energy needs to be released from glucose in muscle cells during exercise. Therefore the heart needs to pump more blood to the cells because the blood will transport glucose and oxygen for respiration to take place.

7 A.

8 a i 4.5 × 10^{12} cells per dm^3.
     ii 8.2 × 10^{12} cells per dm^3.
  b Altitude.
  c If climbers spend time at high altitude their bodies will produce more red blood cells for carrying oxygen to their cells.
  d If acclimatization is not carried out, the number of red cells will be insufficient to carry the limited oxygen that is available. The body will therefore suffer oxygen shortage and the climber will faint because of the lack of energy.
  e The white cells are not needed for oxygen transport.

9 a (i) 0.8 second. (ii) 0.1 second. (iii) 0.3 second.
  b Their muscles relax, the volume increases and the pressure decreases.
  c 0.5 second.
  d 75.

10 a (i) B – C. (ii) C – R. (iii) C – D. (iv) A – B.
   b 12 units.
   c The relative thickness of the wall in the aorta resists pressure changes.

Chapter 3.5 (p. 170)

1 a Silk Cut Extra Mild.
   b i The filter reduces the tar content.
     ii It also reduces the nicotine content.
   c 20 mg/cigarette.
   d It prevents the cilia from carrying out their usual function of cleaning the respiratory tract.
   e Bronchitis, lung cancer.

2 a The volume inside the bell jar increases. The pressure decreases so that it is less than atmospheric pressure. The atmospheric pressure then forces air through the tube into the balloons.
   b (1) trachea; (2) bronchus; (3) lungs; (4) diaphragm

3 a $7dm^3$
   b $37dm^3$
   c Exercise has stopped so there is less need for so much oxygen to release energy from glucose in the muscle tissue. The intake of air therefore becomes less.
   d i To release energy from glucose during respiration.
     ii More energy is needed during exercise. Therefore more air containing oxygen needs to be breathed in.
     iii It increases the rate of breathing and the depth of breathing.

Chapter 3.6 (p. 180)

1 a A Loop of Henle. B Bowman's capsule. C Glomerulus. D Collecting duct. E Blood capillaries.
   b From C into D.
   c Urea, water, mineral salts, glucose.
   d Protein.
   e Glucose.
   f Urea.
   g The ureter via the pyramid of the kidney.
   h Urine.
   i Urea, mineral salts.

2 a Water $1.8\,dm^3$. Sodium 13g. Glucose 0g. Calcium 0.2g. Urea 30g. Potassium 2g.
   b The first (proximal) convulated tubule.
   c Active transport requires release of energy from glucose to pump material against a gradient of concentration. Passive absorption relies on the physical process of diffusion.
   d Osmo-regulation – an active process.
   e The anti-diuretic hormone.
   f The hypothalamus.
   g (i) Urea. (ii) Glucose. (iii) Calcium.

3 a A greater volume of urine was produced.
   b Water was reabsorbed by the kidney tubules.
   c Urine formation takes a longer time in S.

   d There would be more sodium chloride in the urine from S than from W.

4 a The glomerulus attached to the Bowman's capsule.
   b The surrounding fluid is continuously replaced.
   c Protein – this would increase the amount of urea formed.
   Salt – the damaged kidney would be unable to regulate its concentration in the blood.
   Liquids – the damaged kidney would not be able to regulate the amount of water.

Chapter 3.7 (p. 200)

1 a (i) Optic nerve; (ii) spinal cord; (iii) retina.
   b A sensory neurone passes impulses from the receptor to the central nervous system. A motor neurone passes impulses from the central nervous system to the effector.
   c The increase in pressure caused by blowing your nose can be transmitted from the throat through the Eustachian tube to the middle ear. This could push the artificial ear drum out of place.

2 a The brain and spinal cord.
   b The epidural injection contains pain killers which affect the receptors and the central nervous system. It anaesthetises the pain centres of the brain but does not affect the autonomic system which supplies the uterus.

3 B G E D F A C.

4 a (i) 0.05 second. (ii) 0.30 second.
   b The reaction time becomes faster probably due to practice but after 22 attempts the reaction time began to become slower probably due to fatigue.
   c Fatigue has begun.
   d The reaction time would have reverted to that at the beginning of the experiment.
   e Alcohol would probably slow down the reaction time to a value that was lower than at the beginning of the experiment, due to its effect on the cerebellum, with resulting loss of muscular coordination.

5 a 22–25.
   b Methadone.
   c They decrease the basic body metabolism, slowing the heartbeat and rate of breathing.
   d In the stomach.
   e The heart, liver and brain.
   f Nicotine and tar.
   g Nicotine. It affects the cilia by preventing them working to clean the breathing system.
   h The Government health warning on packets of cigarettes. Banning advertising on television.

6 a Addiction when the body metabolism is altered so that it needs drugs to function.
b Slowing the rate of heartbeat. Slowing the rate of breathing. Reduction of body temperature.
c Aspirin.
d Damage to the stomach lining. Irregular heartbeat.
e As a pain killer.
f It is found only in minute amounts in tea and coffee.
g Addiction.
h Alcohol affects the brain giving loss of muscular coordination and balance.
i Alcohol reduces blood flow to the kidneys and therefore reduces their filtration rate.
j It causes more blood to reach the skin capillaries and radiate heat away from the body.

Chapter 4.1 (p. 215)

1 a Mr Smith, Bb. Mrs Smith, bb
John, Bb. Susan, Bb. Peter, bb. Debra, bb.
b John and Susan.
c If Mr Smith had two genes for black hair, all the children would have had black hair.
d 1 in 2.
e Sex. Blood group.
f A mutation is an abnormality in one or more genes or chromosomes leading to the development of an abnormal characteristic of an individual.
g Albinism. Sickle-cell anaemia. Down's syndrome. Cystic fibrosis. Phenylketonuria. Colour blindness.

2 a (i) Cousins. (ii) Joan is Charles' aunt.
b (i) Rh^-Rh^-. (ii) Rh^+Rh^-
c As Rh^- is recessive, the only possible genotype for a Rh^- person will be two Rh^- genes. Charles' wife is Rh^- but Charles is Rh^+. Their daughter is Rh^-, so Charles must be heterozygous for the Rhesus factor.
d No, because Ralph is homozygous for Rh^+ and Jennifer is heterozygous.

3 i Sperms, X^h and Y. Eggs, X^H and X^h.
ii XX, XX, XY, XY.
iii X^hX^H, X^hX^h, X^HY, X^hY.
iv Female, Female, Male, Male.
v Normal clotting, Haemophiliac, Normal clotting, Haemophiliac.

4 a The father.
b Only the females.
c Assume Normal vision = N and blindness = n
(i) X^nY; (ii) XY; (iii) X^nX.

5 a i Parents: HbHb and HbHbS

Gametes	Hb	Hb
Hb	HbHb	HbHb
HbS	HbHbS	HbHbS

ii Parents: HbHbS and HbHbS

Gametes	Hb	HbS
Hb	HbHb	HbHbS
HbS	HbHbS	HbSHbS

b i There appears to be a direct relationship between the presence of the malarial parasite and the incidence of the sickle-cell gene. The reason for this is because those who are heterozygous for the sickle-cell gene have an immunity to malaria. The mutant gene is therefore at an advantage in areas where malaria occurs.
ii The distribution of the malarial parasite is linked to the presence of rivers or lakes where the mosquitoes breed.

6 a (i) Mary; (ii) Monica.
b Susan.
c 3.
d Susan.
e Monica.
f 80%.
g Hand span.
h Eye colour and blood group.

7

	Mother		Father	
Genotype	AO		BO	
Gametes	A + O		B + O	
Genotype of children	AB	AO	BO	OO
Blood Group	AB	A	B	O

8 a

Heights range (cm)	150–154	155–159	160–164	165–169	170–174
Numbers	2	4	6	14	8

b See histogram.

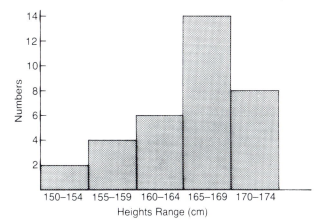

c Weight.
d Probably the most frequent height range would be lower than that for boys.
e At sixteen years of age, boys are at the peak of their growth spurt.

Chapter 4.2 (p. 242)

1 a i Menstruation is the shedding of the inner lining of the uterus periodically at intervals of about 28 days through the vagina.
ii Ovulation is the release of an egg or eggs from the ovary periodically at intervals of about 28 days.
b Six days.
c (i) Oestrogen. (ii) Progesterone.

2 a See histogram.

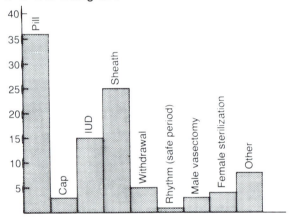

b (i) Pill. (ii) Rhythm (safe period).
c The rhythm method, because there is so much variation in the duration of the menstrual cycle and women cannot be certain that an egg is not present in the fallopian tube for fertilization.

3 a The umbilical vein.
b Carbon dioxide. Urea.
c The heart of the embryo is developed to such an extent that it can pump blood along the umbilical artery.
d The embryo releases energy from glucose for growth.
e It increases the surface area for diffusion.

4 a See graph.
b Non-identical twins are most common in 36 year old women. Non-identical twins are more common than identical twins.
c Identical twins are produced by the fertilization of a single egg and subsequent cleavage of the egg into two separate parts, each of which develops into a baby.
d Non-identical twins can be of different sex because they are formed when two eggs are fertilized separately. An X or Y sperm can fuse with the egg.

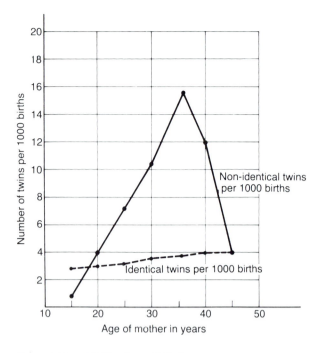

5 a A Head B Nucleus C Tail
b Nucleus in the head.
c Locomotion, so that the sperm can swim to the egg.
d Fertilization.
e Gestation or pregnancy.
f A gamete's nucleus has half the normal chromosome number of a normal body cell.
g X.
h Human milk will contain antibodies for the baby's protection. A closer relationship develops between the mother and baby.
i The Rubella virus can damage the nervous system of the embryo during pregnancy. It is therefore important to protect girls against the virus before they reach puberty.
j If the mother suffers from anaemia it is dangerous for the unborn baby because the baby depends on the mother's blood cells for carrying it enough oxygen for proper development. If it does not get enough oxygen it becomes stunted physically and mentally.
k Iron.

6 a (i) 1075 000 per year. (ii) 650 000 per year.
b (i) 700 000 per year. (ii) 700 000 per year.
c 1922.
d 1916 and 1940.
e An increase. The birth rate greatly exceeded the death rate.
f Child welfare services should find it easier to cope with the falling child population. The Health services would find it easier to cope with maternity care.
g Better diet, better medical care, better health education.
h Concraception means the prevention of sperms reaching eggs. Pill, sheath, IUD, Cap.

7 a Increased promiscuity. More people who are infected going to clinics for treatment.
  b Between 1971 and 1978. 168 804 new cases were reported.
  c Better health education to make people aware of the dangers. The more widespread use of male condom-type contraceptives.
  d Males : pain on passing urine. Discharge from the urethra. Abscesses in the urethra. Females : pain on passing urine. Vaginal discharge and infection of the rectum.
  e Penicillin is the most widely used antibiotic despite the percentage of resistant strains of bacteria being up to 80 in some parts of the world. Streptomycin and tetracycline are antibiotics that are also used.

8 a Sperm cannot reach the egg for fertilization.
  b i Usually one egg is produced per month. This would be too difficult to find. It increases the chance of success when more are fertilized.
   ii Eggs are at the right stage of development and are ready to be fertilized. They are near to the surface of the ovary and are easier to collect. Approximately days 12–14.
   iii It must contain food (glucose), oxygen. It must be well aerated, at body temperature and sterile. It must be kept in the dark to simulate the conditions within the body.
   iv The doctor must wait until the uterus is ready to receive the embryo. It takes three days for an embryo to reach the uterus normally. It will not implant until several cells are formed. The doctor can check that embryos are developing normally.
   v The cervix is the natural opening. There is no damage to the uterus. The ovaries cannot be reached by the natural opening. The embryo is placed where it would arrive naturally.

  c No test tube is involved in the process at any stage. The embryo only grows to the 16 cell stage – it is not a baby.
  d i The chromosomes can be studied and the X or Y or both can be identified.
   ii An abnormal number of chromosomes can be counted.
  e i The amniotic fluid and placenta.
   ii More materials must be transported to and from the developing embryo by the blood stream.
  f i They contract.
   ii They relax.

9 a A XX. B XY. C XX. D YY. E Y. F X. G X. H X. I XY. J XX. K XY. L XX.
  b Meiosis.
  c Fertilization.

10 a pH, dilution of sperms, carbon dioxide, time limit for life of spermatozoa.
  b Sperms die in acid conditions. In a large volume of semen the sperms may be too diluted to have sufficient density for fertilization. Too much carbon dioxide inactivates sperms. If too little carbon dioxide is present, sperms become too active and exhaust their store of energy. Sperms only retain their activity for forty-eight hours.

11 a Between the 9th and 10th lunar month.
  b 180 mm.
  c The total length of the foetus (at 6 months).
  d 3 : 2.
  e Between the 4th and 8th lunar months.
  f Between the 8th and 10th lunar months birth occurs. The passage of a larger head and trunk through the cervix and vagina would be more difficult if the head and trunk continued to grow at the former rate.

Index